Praise for *Wi-Fi and the Bad Boys of Radio*

"After bringing modern communications to Alaska's Native villages, Dr. Alex Hills continued on to make great contributions in the world of wireless technology. Dr. Hills is a fine writer and teacher, so I have no doubt that his book will be both fascinating and entertaining."

— Walter J. Hickel, (1919-2010),
former United States Secretary of Interior

"I know of no one so capable of telling the Wi-Fi story and explaining so clearly how the technology works. Alex Hills is certain to capture the public imagination with this new book."

—Jim Geier, Principal Consultant, Wireless-Nets, Ltd. and Wi-Fi author

"Alex Hills has contributed to the developing world and to developing advanced wireless technology at one of the world's most tech-savvy universities. Working on both frontiers, Dr. Hills pioneered wireless Internet and launched a revolution in the way the world communicates. His story of how we "cut the cord" begins in a place where there were no cords to begin with — remote Alaska."

— Mead Treadwell, Lieutenant Governor of Alaska and
former Chair, United States Arctic Research Commission

"Being from Alaska, I am aware of the great contributions Dr Alex Hills made to my state in building its communication systems. Later, I discovered the importance of his Wi-Fi work through an article about him in The Economist. Alex's work has raised the quality of a lot of people's lives, including mine."

— Steve Cowper, former Governor of Alaska

Wi-Fi
and the
Bad Boys of Radio

Dawn of a Wireless Technology

Alex Hills

First published by Dog Ear Publishing
4010 W. 86th Street, Ste H
Indianapolis, IN 46268
www.dogearpublishing.net

ISBN: 978-145750-560-7
Library of Congress Control Number: 2011914066

This book is printed on acid-free paper.

Printed in the United States of America

For Meg, Becky and Karen

CONTENTS

FOREWORD

Danger, opportunity, and freedom are found on the frontier. So are courage, curiosity, and creativity in the pioneers who push the frontier ever outward. At the cutting edge of human activity, it is not protecting the status quo that matters – it is getting things done.

Freedom to think means freedom to innovate.

Alaska is America's frontier, and its imperatives – survival of its people, health care for rural areas, fishing in oceans that can breed the "mother of storms," and finding energy sources and minerals in the Arctic – have brought us major advances in the way we communicate, map the world, and navigate. Even with nineteenth century conditions in some areas, Alaska still is a proving ground for twenty-first century technology. Alaskans are both pioneers and innovators.

And Dr. Alex Hills is no exception. After working to bring telecommunications services to remote parts of Alaska, he went on to become a Wi-Fi pioneer – with global impact.

As I write this aboard a 737 at 36,000 feet, I am connected to the Internet through an onboard Wi-Fi system. At my home, my office, a hotel in New York, or an airport in Tokyo, I use Wi-Fi to stay in touch via email, voice and video. Wherever I go, I am "wired in" because the world is now unwired. Wi-Fi connects us, whether across the last mile or the last few feet – wirelessly.

Alex Hills has made contributions to developing nations around the world, and he has also worked at one of the world's most tech-savvy universities, Carnegie Mellon in Pittsburgh. Working on both geographic and technological frontiers, Dr. Hills pioneered wireless connectivity to the Internet with the spirit of a roamer – someone whose theme song might be "Don't Fence Me In." For roamers among the rest of us, those who need to be in touch as we move freely across campus or across town, we have Alex to thank for launching a revolution in the way the world communicates.

I have had the opportunity to work with Alex in government, business, and academia. I am delighted we have this log of his explorations and his trail markers to help us continue to pioneer and innovate.

—Mead Treadwell, Lieutenant Governor of Alaska;
former chair, U.S. Arctic Research Commission;
former chair and CEO, Venture Ad Astra.
June 2011

ACKNOWLEDGMENTS

Some have called me the inventor of Wi-Fi. I beg to differ.

Like most modern innovations, Wi-Fi resulted from the work of many, some of whom appear in these pages. This is their story as much as my own. And there were many others – working in universities, companies and laboratories outside Carnegie Mellon – who contributed to the development of Wi-Fi. Their absence from the narrative certainly does not diminish the importance of their work.

I'm indebted to those who patiently reviewed drafts, provided advice, checked facts, and gave encouragement. They are: Robert Burns, Nellie Moore, John Lee, Dan Siewiorek, David Johnson, Mahadev Satyanarayanan, Mark Campasano, Cees Links, Lisa Picone, Bruce Tuch, Bob Friday, Paul Dietrich, Jon Schlegel, Rebecca Hills, Karen Hills, Richard Chiappone, and David Cheezem. I gratefully share with them the credit for whatever came out well – but none of the blame for what didn't.

But my greatest debt is to the beautiful young nurse I found in remote Alaska. Over the years, my wife Meg has put up with a lot, including the writing of this book. I'm deeply grateful to her.

—Alex Hills, Distinguished Service Professor,
Carnegie Mellon University
June 2011

CHAPTER 1

DISCOVERING RADIO

JULY 2010

It's a small town. It could be anywhere in the American Midwest. There are no traffic lights in the town center – only a flashing red light at the main intersection, where a farm tractor halts and then crawls forward. Radio waves swirl everywhere, but they are invisible.

Each day the little hamlet wakes up, rubs its eyes, and begins to move. One by one, cars stop at the blinking light. They pause and then proceed deliberately along Main Street – with extra care when children are nearby. People drift in and out of the Main Street shops, stopping to chat with friends and neighbors. A raised voice is rarely heard. It's a friendly little place, and, except for the occasional political dispute, everyone gets along.

But technology has found its way to our little town. Shoppers have cell phones pressed to their ears. Drivers chat with invisible companions. A young couple sitting on a park bench beside the train depot uses their smartphone to check e-mail. Though they're unaware of it, they're connected to a wireless hotspot installed in the town center by a local civic group. There is free Wi-Fi service for all.

The pair of 20-somethings shows their new device to a passerby. Folks like to be connected, having family and friends always available and Internet information instantly accessible. Young people say wireless is "cool." Their elders say it's "handy."

Like my friends and neighbors, I use my smartphone every day. It's a cellular phone, e-mail reader, calculator, Web browser, music player, newspaper reader, book reader, calendar, camera, notepad, and more. Wireless technology and small computer-like devices have become part of our lives.

But, when my neighbors hear that I work with wireless, they ask questions. Why don't smartphones always work? Why do Blackberrys sometimes garble voices or go completely dead, making it impossible to place a phone call or read e-mail? Why does an iPhone work perfectly in one corner of a room but not another?

I search for simple answers to my friends' questions. Radio waves are the bad boys that stir up the trouble, but their mischief is subtle and complex. They act up in endlessly creative ways. Their misbehavior is diabolical. Designers of wireless devices scramble to outwit the impish waves but never claim total victory.

As I struggle to answer my friends' seemingly simple questions, I flash back to scenes of my journey in the world of wireless. I remember vignettes from a lifetime searching for answers to its mysteries. But the answers to the questions depend on which of the bad boys is misbehaving. I leave the young couple, walking around a building's corner and into its parking lot. My e-mail service stops working. Yet the Wi-Fi station is only a few blocks away. What's the problem?

That night I drive my pickup out the highway north of town, and the truck's AM radio pulls in a West Coast station. I hear other faraway stations fade in and out, crowding out the local ones. Sometimes I hear two voices at the same time, one from nearby and one from a few thousand miles away. A faraway station overpowers a closer one, allowing only the distant one to be heard clearly. Then the distant signal fades out, and I hear the closer one again. This never happens in the daytime. It happens only at night. Why?

I switch to FM, and a local station flutters in and out as I continue to drive. How can this be? The station's transmitter is only a few miles away. And I don't hear any distant stations on FM. Why not?

Even the cell phone service is spotty. I climb a hill, and my cell phone works, but, when I crest the hill and descend the other side, the connection is lost. Other times the cell phone's sound quality deteriorates from bell-like clarity to noise-corrupted chaos in just a few moments. Then, suddenly, it's clear again.

When I read e-mail messages on my smartphone or my Wi-Fi equipped laptop computer, the devices try to shield me from the shenanigans of radio signals. Because I don't hear any radio garble, the problems are hidden. Sometimes an e-mail message or Web page appears on my screen and sometimes it doesn't, but the underlying reasons are invisible.

My life has been a quest – a mission to understand the strange behavior of radio waves. I've hacked my way through the dense jungle of AM broadcast reception, trying to pick out and separate one signal from another. I've wandered in the swamps of ham radio and short waves, where things were even messier and more disorienting. And I've trekked across the arid deserts of cellular and Wi-Fi, searching for high terrain from which to survey my surroundings and pick up elusive signals.

Throughout the journey, I looked for technological tools to help me navigate – tools that could separate one signal from another and help to provide reliable communication service. I needed ways to beat the bad boys at their own game. I searched for ways to get the message through.

My journey in the world of wireless began when I was a teenager – more than 50 years ago. In those days we just called it radio.

((•))

A yellow-orange glow filled the room. Created by the vacuum tubes of the receiver and transmitter, the glow lit up the entire ham shack. It was late at night in the fall of 1958, and I was "pounding brass" – using international Morse code to communicate with a fellow ham operator half a world away.

My forefinger and thumb nudged the telegraph key's paddle, first right, then left. Tiny sparks jumped between the key's contacts. The yellow-orange glow from the big radio receiver's tuning dial illuminated the paper I used to copy down my faraway friend's message.

The ham shack wasn't really a shack. It was a corner of my third-floor attic bedroom, where radio equipment was stacked on a shelf with a tangle of wires and cables tying together the pieces. One cable went through the wall to a wire antenna I had strung between two trees outside my bedroom window. The confinement of the ham shack intensified my connection to the receiving and transmitting equipment. I could smell the dust that had collected inside the equipment cabinets, and I could feel the heat radiating from the vacuum tubes that lit up the room. But the closeness of the physical space contrasted with the vastness of a virtual space – that of my fellow ham operators. The ham shack was my connection to an exotic outside world.

The big radio receiver was my pride and joy. I had bought it secondhand and souped it up by substituting vacuum tubes more sensitive – I called them "hotter" – than the ones originally installed.

I was a happy teenager as I used a special telegraph key – one that could send international Morse at high speeds. I had abandoned a traditional telegraph key in favor of this one. The newer key, called a "bug," spewed out a blur of *dits* and *dahs* – "dots" and "dashes" to non-telegraphers – that were my message to a faraway friend. International Morse code was the best way to communicate under difficult radio conditions. It was better than voice communication because it was possible, by using electronic tricks and a radio operator's sensitive ear, to get a message through. Even though my transmitter wasn't very powerful, its signal was able to punch through the static when I used Morse code.

I had at first used a traditional telegraph key. I bought it on sale at a military surplus store in nearby New York City, where I found it tucked beside a stack of dusty old World War II radio transmitters. This is the kind of key you see in the movies being used by a wartime shipboard radio operator or a Wells Fargo telegraph operator in the Old West. It had a knob meant to be held between thumb and forefinger. The motion was up-and-down. I pressed down and then released the knob, opening and closing a single pair of electrical contacts, a faint spark sometimes visible.

A quick press down gave a *dit* and a longer one gave a *dah*. It took some skill to use the key to produce a string of *dits* and *dahs* intelligible to another operator – but not as much skill as I needed to use the bug.

Technically called a semiautomatic key, the Vibroplex bug allowed me to send international Morse at high speed. I had diligently saved my money in the hope that I could someday move up to a bug, often eyeing the shiny gadgets in the radio catalogs stacked in my ham shack. With a bug the operator holds a small paddle between thumb and forefinger, and the motion is not up and down but sideways – to the left and to the right. A push to the left is similar to pushing down with a traditional key and is used to send the longer *dahs*. Pushing the paddle to the right sets a metal reed vibrating, and an electrical contact near the end of the reed touches another contact, issuing a string of short *dits* for as long as the operator holds the paddle in this position. The vibrating reed inspired the bug's manufacturer to name the company and its product "Vibroplex."

The Vibroplex bug produced impressive results. As my fingers flew and the reed's vibration stopped and started, a fast, well-formed string of *dits* and *dahs* sped across the airwaves.

After finishing my transmission, I flicked on the big receiver, and my faraway friend began to send. His signal was weak, and there was competing static that made it difficult to hear. I reached over to the receiver and switched on a filter that made his signal louder, bringing it up and out of the noise. At the same time, the filter caused an eerie ringing sound. Each time the distant operator released his key at the end of a *dit* or *dah*, the tone persisted, fading out gradually.

Copying Morse code through noisy and difficult radio conditions was part of the challenge of being a radio ham. Receiving the message, even through poor atmospheric conditions, was by the 1950s a part of the radio tradition. For decades, other radio operators had done the same, sending and receiving important messages in times of war and peace. My use of radio was only a hobby, but I knew I was continuing the tradition of the many radio operators who had come before. Now, more than 50 years later, radio has become an unseen yet important part of everyday life, and the telegraph key has been replaced by digital technology.

The smartphone I now carry, like my teenage ham radio station, depends on the successful transmission and reception of radio signals. Unlike the ham radio signals, they travel shorter distances and they're more likely to be blocked by obstacles. But they're still radio signals.

Radio – wireless – is a funny thing. Its signals can misbehave without warning. Sometimes they successfully reach your cell phone, Android phone, Wi-Fi equipped computer, or AM/FM radio. And sometimes – more often than you might like – there are problems. It can be inconvenient and frustrating as these personal devices become commonplace and seemingly indispensable.

Most users of smartphones and cell phones don't think about the vagaries of radio except when there are problems. I'm a little different – my journey has sensitized me to the problems of coverage and reception. I programmed my cell phone to read the strength of cellular signals. It's a more accurate indication of radio coverage than "can you hear me now?"

or even the familiar one to five bars. It would be hard to prove that this numerical readout actually helps me better use my cell phone, but old habits die hard. I just like to know the quality of the signal I'm using.

Modern cellular phones and smartphones – Android phones, iPhones and Blackberrys – are designed to shelter you from the difficulties of radio reception. Unlike older analog devices, they do not alert their users with noise or low quality sound when signals get weaker. Instead, they keep problems hidden until the received signal is completely useless. When this happens, service can be suddenly interrupted and the connection lost.

Owners of digital TV sets who use them to pick up over-the-air TV signals notice the same effect. There is no snowy reception – instead, either crystal-clear pictures or nothing. There is no audible noise – just crisp sound or nothing. The technology of digital communication has this all-or-nothing quality.

But the equipment in my ham shack was analog. I could hear my ham radio friends' signals fading in and out. The noise level rose and fell, sometimes overriding the signals. I had to work to get the message through.

To qualify for my first ham radio – officially amateur radio – license at the young age of 14, I was required to send and receive – radio operators said "copy" – international Morse at the slow speed of five words per minute. There was also a written test.

On April 12, 1957, the mailman delivered a small envelope to my family's suburban New Jersey house. The envelope showed the return address of the Federal Communications Commission – the FCC. Now, more than 50 years later, I still remember the date. I had taken the test for my Novice Class amateur radio license a few months earlier. I knew I had passed the international Morse test, but I wasn't sure about the written test.

I had walked the half-mile from school to our leafy neighborhood in the New York City suburb of Caldwell, New Jersey. On that sunny afternoon, I wondered if this would be the big day. And the small envelope I found in the mailbox quickly transformed a routine day to a memorable one.

The envelope had a musty smell – maybe from the years it spent in an FCC storeroom – and my hand shook as I opened it. But I relaxed when I saw that the envelope contained an amateur radio license with call sign KN2ZMO.

I ran up two flights of stairs to the ham shack. The afternoon sun made it hot up there. The air smelled stale. I hurried to open the windows to get some air moving. I flipped switches on both the receiver and transmitter. The transmitter had been used only once before – when I asked a friend with a ham license to test it for me.

My first radio contact came quickly. Tapping out Morse code on the military surplus key – haltingly using the traditional up and down motion – I made contact with another novice ham in Wooster, Ohio, more than 400 miles away. His Morse signal was weak, but I could faintly hear it through the noise. Listening intently, I wrote deliberately with a pencil. "…R-S-T….5-7-9…" ("Your signal has excellent readability, good strength, and excellent tone.") "QTH WOOSTER OHIO…" ("My location is Wooster, Ohio…")

Like some ham radio exchanges, this one was impersonal – not much more than an exchange of signal information and locations. I was too excited to send more. But it was my first contact as a ham radio operator! And two states away from New Jersey, too! Soon I had made contact with other novices across the United States, and a few months later with hams in Libya, England, Germany, and other countries around the world.

The third floor ham shack was my connection to a secret world unknown to most of my high school classmates – a world of radio waves and faraway places. Sometimes I tried to explain it to my friends as we walked between classes or stood around the school yard, but most of them cared little and understood less about my other life. To them, lettering in a sport or participating in a school activity was what really mattered.

Usually I walked home from school in the company of friends Ray, Bob and Greg, a trio who understood the secret world. They were also radio hams, each with his own ham shack and call sign.

We would have been "geeks" if it had been a word in the 1950s. Ray was round, Bob slender, and Greg muscular and energetic, with the swagger that came from being two years older than the rest of us. As we walked, we talked urgently about our latest radio construction projects, faraway radio contacts, or taking the FCC exam for a higher level ham license. Other teenagers couldn't understand our excited conversations laced with ham radio jargon. QSO meant radio contact. QTH was the location of the ham station. QRM was radio interference. To others we were speaking a foreign language. But we didn't care what they thought. Ray, Bob, Greg and I lived in a different world – the world of radio.

Ray and I developed a special bond. Our walks home from school often led to my ham shack or his, where we worked together on an equipment building project or diagnosing a technical problem, working together to brainstorm a solution. At night Ray and I talked endlessly by radio or on the "landline" – a word that was then a technical term and had not yet crept into common use. Our nighttime conversations were sprinkled with QSO, QTH and QRM, as we chattered in our special language.

Ray, Bob, Greg and I were not completely without social skills. We were in clubs and the high school band, and I was a manager of the football team. But we knew what was really important. We had our priorities, and they were not school activities or classes.

My Novice Class license was a beginner's license, requiring Morse proficiency at only five words per minute and a little knowledge of radio theory and FCC rules and regulations. The license was valid for one year, allowing enough time to gain the experience and ability needed to qualify for a higher grade license. I did not disappoint the FCC. In less than a year I had increased my code speed enough to pass the Morse test at 13 words per minute and studied to pass a more difficult written exam to qualify for a General Class radio license with the call sign K2ZMO. The N was dropped because I was no longer a Novice.

Learning to send and receive international Morse was like learning a language. In order to do it well, I had to master the sound of Morse characters – not an abstract notion of dots and dashes. A Morse "a" is sometimes visually represented by a dot followed by a dash (. _). This can be translated to a short sound followed by a longer sound. But to a radio operator a Morse "a" is a single sound: *didah*. This particular sound is instantly interpreted by the operator's brain as the letter "a." No conscious thought is needed. Hearing *didah* and thinking "a" is more like a reflex. When using a key, the telegraph operator sees an "a" on the page but hears and then sends the sound *didah*. Young people are adept at learning new languages, and this is true even when the new language is Morse code.

Morse is sometimes called an *efficient* code because it allows a message to be sent in less time than other similar codes. The letter "e" could have been represented by *dahdahdidah*, but the code's designer chose to use *dit*. He chose symbols so that the most common English letters were represented by the shortest symbols. This is what allows messages to be sent so quickly. The most common letter "e" is represented by *dit*, the shortest symbol, and the second most common letter "t" is represented by the second shortest symbol, *dah*.

The legend is that Samuel Morse's associate went to a printer's shop and looked at the bins where the type for each letter was stored. The bin for "e" was stacked highest with type, and the bin for "t" was second highest. These bins full of type told him how common were "e," "t," and all the other letters. This trip to the printer's shop helped to design an efficient code. Later mathematical analysis showed that the symbol lengths were almost exactly right.

Radio telegraphy was an all-or-nothing technology. It was either on or off at any instant. There was no in between. When I pushed down the key, I turned the transmitter on, sending out a signal. When I released the key, I turned the transmitter off. A radio receiver tuned to a Morse signal produced a tone when the signal was received. Otherwise, there was no tone. When an operator sent a Morse "a" (*didah*), the receiving operator heard a short tone followed by a long tone. The conversion to text happened inside the receiving operator's head. The "signal processing" was done not by electronic circuitry but by a human computer.

A skilled Morse operator was said to have a good "fist" – the ability to send with a key, converting a string of text to a well formed string of *dits* and *dahs*. Like the ability to speak clearly in a foreign language, copying the accent of a native speaker, it was a matter of hearing and imitating sounds. A highly skilled operator could recognize the sounds of complete words, sending and receiving messages at about the same speed you can speak and listen in English. Some operators could listen to and understand a Morse message sent faster than they could write it. They didn't need to write anything at all. And still others used a typewriter – a "mill," to a radio operator – faster than pencil and paper.

Developing Morse skills was time-consuming, even for a young learner. I spent hours in the third-floor ham shack listening to the Morse code practice transmissions of W1AW, the official station of the national ham radio club called the American Radio Relay League. The station sent code practice transmissions every night.

I also practiced by listening to phonograph records containing Morse code and by using a special paper tape code practice machine. The contraption, incorporating the same principle as a player piano, used rolls of paper tape with a series of holes that looked like dots and dashes. The paper tape ran through a pair of contacts, something like the contacts of a traditional telegraph key, and the contacts turned on and off a tone generator that produced the familiar *dit*s and *dah*s.

As I increased my skill level, I was able to communicate with ham radio's big guns, those who operated at high speeds and used the radio frequencies that were, in effect, reserved for them. Most were older than me.

One experienced operator was Ed Raser, a ham radio elder who commanded great respect. Using the call sign W2ZI, he had been operating from Trenton, New Jersey for a long time. When I first heard him on the air, I knew immediately that his call sign – with only two letters and not the usual three after the numeral 2 – was a very old one. Ed was a ham radio pioneer who had been using the airwaves since the dawn of radio – the earliest years of the twentieth century. He sent his first radio transmission in 1912 – 45 years earlier than my own awkward contact with that Wooster, Ohio ham. His call sign had previously been W3ZI, before the "3" was changed to a "2," and before that it was merely 3ZI. The W prefix was added to indicate a radio station located in the United States.

I sometimes talked with him using voice transmission, each of us using a microphone instead of a key. Ed's voice was gravelly – he sounded like an old-timer – but gentle. He showed others how to be a capable but courteous ham.

Over the decades Ed had collected lots of ham equipment. His house in Trenton was a virtual radio museum that documented the growth of radio technology since its birth. Eventually, Ed made his collection a real radio museum, and it attracted lots of ham visitors to Trenton. I talked with Ed by radio often but met him face-to-face only once – at a ham radio gathering in Browns Mills, New Jersey. The white-haired gentleman from Trenton was godlike to me. Our overlapping years were few, and I was lucky to have known him.

I qualified for the highest amateur radio license, Amateur Extra Class, which required sending and receiving at 20 words per minute. And, like other hams, I was proud of my international Morse ability. I received a certification by the American Radio Relay League that I was proficient at the blazing – to me – speed of 30 words per minute. But there were more experienced operators who could easily do speeds as high as 40 or 50 words per minute.

Morse code skill was a badge of honor among radio operators in the 1950s. Even as technology improved and allowed the use of robust "single sideband" voice signals, Morse code still punched through the most difficult radio conditions of interference and static. Morse skill eluded many, and those who were proficient were an elite group within the ham radio fraternity.

Years later, in 2006, the FCC announced that the Morse code test would no longer be required to qualify for an amateur radio license. The commission said that technology had advanced so far that there was no need for people to use the code. This was an understandable conclusion, but some traditionalists disagreed. They saw the decision as an unfortunate and misguided one. Old-timers continued to be proud of their ability to send and receive international Morse. Some continued to cling to the image of an emergency situation where the only way to communicate might be by touching together two wires sticking out of a transmitter, switching the transmitter on and off to send a distress message.

I was torn. The commission was correct in its reasoning, but I was by this time an old-timer too. I shared the nostalgia for international Morse. Yet I recognized that technology had moved on. The FCC decision had brought the end of an era. But the commission's decision was really just a belated acknowledgment that digital technology had made radio telegraphy – like wire telegraphy – technologically obsolete. Ships at sea and explorers in remote lands can now use digital systems – like satellites that provide coverage of the whole world – to communicate. No longer do we need to touch two wires together in an emergency, tapping out *dididit dahdahdah dididit* (S-O-S).

Yet telegraphy was, in a way, the first digital communication technology. Like modern digital systems, telegraphy embodied the idea of "all on" or "all off," with nothing in between. The telegraph key was either pressed down or it wasn't. The current either flowed or it didn't. Modern digital technology is the same. Only two signals are possible, a "one" – current flows – or a "zero" – current doesn't flow. With Morse code radio telegraphy, it was left to the radio operator to decide whether the distant key had been pressed and to translate its "ons" and "offs" to text. Today's computer-like circuits automatically receive and interpret digital signals.

Technology has changed, but the underlying radio signals haven't. There are still transmitters and receivers, though they're embedded inside your cell phone or smartphone. Transmitted signals must still travel to the receiver's location, and they must be strong enough to be understood by the receiver circuitry that has replaced the radio operator's ability to interpret signals. The laws of physics governing radio signals' behavior are the same. Radio's behavior fascinated me as a young radio ham. Now, more than 50 years later, it still does.

As a young ham operator in the 1950s, I built my first ham transmitter, starting with nothing more than a metal chassis and a handful of parts provided by a kit manufacturer. Working in the ham shack, I first mounted the larger parts – vacuum tube sockets and terminal strips – using screws and nuts to hold them to the chassis. The next step was to add wires and small parts of many colors, wrapping their ends around the appropriate terminals.

I upended a bag of small parts, which spilled onto the workbench, a random jumble of color – bright reds, yellows, greens and blues. The parts included resistors – small cylinders with wires extending from their ends and trimmed in colored stripes, each stripe with a meaning. There were also capacitors – discs with two wires protruding and stamped with colors or numbers that gave their electrical characteristics – and coils of wire, called inductors, with a few or hundreds of turns. My task was to organize and interconnect this jumble of parts. The finished product, a piece of radio equipment, might have been seen as an artistic creation, but the pieces were carefully selected and wired together to produce an intelligible radio signal – not a piece of art.

Each color had a meaning. I had learned that brown signified the numeral one, red represented two, orange was three, and the seven other colors rounded out the ten numerals of radio's "color code." The colored bands on a cylindrical resistor were not for aesthetic effect. They were to represent a precise electrical value, its resistance. But the array of bright hues was still beguiling.

The heady smell of rosin filled the air as I began to solder the transmitter's connections. The solder melted and flowed around each clump of wires and then cooled and solidified to form a strong and dependable joint. I could feel the intense heat of the iron's tip – even through its insulated handle – as I applied it to each gang of wires. After the wires were searing hot, I touched them with a wire-like piece of solder, a mix of tin, lead and rosin. Tin and lead melted onto the joint, and the rosin sent fumes into the air. I liked the rosin's scent. It signaled progress and the imminent success of the project.

My excitement became nervousness when I first turned on the new transmitter's power switch. This was the moment of truth. I knew that, if I had made a wiring mistake, there could be disaster, with black smoke filling the room and the destruction of the transmitter's components. So I checked my work carefully before beginning the "smoke test." If everything worked as planned, my first transmitter would soon become part of the ham shack's equipment lineup.

In those days it was common for radio hams to build their own equipment, especially transmitters. I built mine from do-it-yourself kits. Heathkits, the most popular ones, were sold by the then-famous Heath Company of Benton Harbor, Michigan. And the most powerful Heath transmitter was commonly called a "Benton Harbor kilowatt," though its actual power was far less than 1000 watts.

But a kit was the electronic equivalent of paint by number. Highly skilled hams often bypassed kits and built their equipment from scratch – some using their own designs. These smart guys first sketched out a "schematic diagram" on paper and ordered the parts that would be needed. Later they used special hole punches to make needed openings in a metal chassis, wired in a rainbow array of small parts, and heated up a soldering iron to finish the job.

My home-built transmitter, with its low-power signal, could be heard halfway around the world. I wondered how this could happen. The signal emitted by the little transmitter was weak, and it became weaker as it traveled farther and farther from my antenna. How could this signal possibly be strong enough to be heard by another ham thousands of miles away? I wondered about this even as I communicated with my faraway friends late at night.

By day I searched out and studied radio theory books that I found at the Caldwell Public Library. My high school homework assignments were a mere annoyance and a distraction from this more important work. I was trying to understand what makes radio tick. The inner workings of my transmitter and receiver were at first a mystery, but, over time, I began to understand the theory of electronic circuits and the design of transmitters and receivers. But the way that radio waves traveled through space continued to baffle me.

I was often in the ham shack past midnight. As I pushed the Vibroplex paddle back and forth, I began to get a feel for how radio signals behave. My radio friends' signals pierced the static. As I copied down their messages to the light of the radio dials and vacuum tubes, I was starting to notice patterns in radio reception. I knew that some radio frequencies worked well for nearby stations up to a few hundred miles away. But, for more distant stations, thousands of miles away, I needed to flip a switch and twist a knob, bringing in a different frequency band. The situation was different at night than in the daytime. On my favorite bands, transmission distances were usually greater during the late night hours, and that's why I was in my ham shack while the rest of the family was asleep and unaware of the activity in the attic.

I had read that a radio signal was a wave moving through space from one point to another. But what was this wave? Could I see it? What was doing the waving? What was really moving? These questions bothered me, even as I flipped the switches and turned the knobs.

A few years earlier, my father had encouraged my interest in radio, probably thinking it would keep me out of trouble. He was partially correct. But I wondered what radio waves really were.

When I was only ten, Dad gave me an inexpensive "crystal set" – a primitive kind of radio receiver in which a thin wire, called a "cat's whisker," made contact with a quartz crystal. I fiddled with the cat's whisker to bring in New York City's radio station WNEW, and it worked! I could hear Nat King Cole singing the gentle ballad "Unforgettable" – one of the top tunes of 1952 – and then the soothing voice of the disc jockey Martin Block introducing the next tune on his "Make Believe Ballroom" radio show. Receiving these familiar voices seemed a big success, but I still wondered about the mysterious waves arriving at the crystal set.

Dad had an intimidating presence. He was bigger than me and didn't often say much after his day at the office. I hesitantly asked him, "How is a radio signal a wave? What is really waving? Is anything moving?" His terse response was "Well, it's just a wave. That's all I can tell you." This wasn't the answer to satisfy a curious young mind, but his answer was firm and its tone final. I didn't persist, sensing I wouldn't get a better answer. But the question bothered me for a long time.

Years later I realized that a radio wave was something like the disturbance that happens when a pebble is thrown into a pond. The stone hits the surface of the water and causes ripples that move outward in concentric circles. This was a wave I could understand. It was easy to visualize. I imagined taking a snapshot of the ripples in the pond at one instant of time, picturing myself using a special camera that could photograph a cross-section of the water, allowing me to see the shape of its surface. In the photo the water surface looked like a squiggle on the frosting of one of my mother's cakes.

But I could also see the position of the water molecules within this cross-section at each instant of time. A series of such images showed that each water molecule actually moved first up and then down. This up and down motion was not at first obvious because the ripple seemed to be moving horizontally. But, if I traced the motion of an individual molecule, I saw that it was actually moving vertically, even as the wave moved horizontally. My little thought experiment, focusing on an individual water molecule and how it moved over time, showed that the particle's motion was also a wave. I made a graph of a single particle's motion, showing its position at different instants of time. To my amazement, the graph formed a wave that also looked like the squiggle on a cake. The pond ripple was a wave that existed in both space and time.

Next I imagined a series of snapshots taken at regular time intervals. Each snapshot showed a wave that existed in space at one instant of time, but the whole set of snapshots showed that the wave was, over time, moving in space. This was the same horizontal motion I saw as the wave on the pond.

Later I imagined that a friend and I were holding the two ends of a rope. I gave my hand a shake, causing a wave to travel along the rope from my hand to my friend's hand. Successive snapshots of the rope were like successive snapshots of the pond ripple. The rope wave also existed in both time and space. The waves in the pond and the rope were easy to visualize and easy for me to understand. Eventually, I learned that a radio signal is another kind of wave, an electromagnetic wave – involving electricity and magnetism – that exists in both space and time.

Experiencing the behavior of my ham radio signals was one step in answering my questions "what is a radio wave?" and "why do these waves behave so strangely?" But I had much more to learn.

((•))

A few years later I was a shy, lanky, college freshman. The campus of Rensselaer Polytechnic Institute – also called RPI – was a collection of ivy-covered brick buildings high on the hill above Troy, New York. The Rensselaer campus looked like that of a classic New England college, but the city of Troy had been a victim of the departure of the textile and steel industries in the late nineteenth and early twentieth centuries. The tidy, hilltop campus stood above a worn-out city below.

Life at Rensselaer was, in some ways, an extension of my ham radio days. I worked at the student radio station WRPI and there found some kindred spirits – other radio geeks. We operated the little FM radio station from cramped quarters in a building we shared with the Rensselaer drama club. The back-stage area of the club's theater was also the home of our 1000-watt transmitter, enclosed in a metal cage to protect it from the thespians, who might accidentally disrupt our finely tuned controls.

The WRPI studios gave meaning to the word "spartan," its plain walls adorned only by the pocked acoustic tiles that we used for sound deadening. A student engineer huddled on one side of a glass window in a tiny control room crammed with electronic equipment that was all within easy reach. A student announcer sat just a few feet away – separated from the engineer by a double pane glass window – in a bare studio. The station had only one other room, a little record library, where we stored our collection of vinyl LP records.

Most of our audio equipment was homemade. Though you couldn't tell by looking, it was technically superior to the commercially available equipment of the day. And it had been designed and built by students. What our little station lacked in elegant décor and spaciousness was compensated by superior sound quality – "high fidelity" was the word we used.

I soon met Professor Hiram Harris, a radio pioneer who had worked on building Rensselaer's historic first radio station. With call letters WHAZ, the AM station began transmissions in 1922 – ranking with KDKA Pittsburgh and WGY Schenectady as one of the nation's first radio broadcast stations. The venerable professor had been a founder of broadcasting in America. Tall and white-haired, he was both distinguished and practical. He spoke softly, with great authority, and he chose his words carefully. Professor Harris was as comfortable handling a soldering iron as a piece of blackboard chalk. He was about more than just theory. All of us at WRPI revered him.

And he was patient with neophytes like me. After accepting me as his student, the professor taught me practical aspects of antennas and radio waves. His insights were different from the theoretical courses I was taking – they made more sense because they were connected to my own experience. I was beginning to understand why, as a ham operator, I had been able to send and receive signals over such long distances.

I learned in my courses that the theory of radio propagation had been mathematically described by the Englishman James Clerk Maxwell in his 1873 two-volume work titled *A Treatise on Electricity and Magnetism*. Maxwell's electromagnetic theory was elegantly summarized in his four famous equations – Maxwell's equations. The elegance was in the equations' simplicity and their ability to describe a very wide range of electromagnetic phenomena.

But Maxwell's equations alone didn't tell me why radio signals move through the atmosphere differently at one frequency than another. Without other information, the equations would not tell a radio operator why one frequency would be good for communication with Wooster, Ohio, but another would not. Radio operators gain this kind of practical knowledge by "feel" – learning through a thousand small experiments which frequency works best. Professor Harris gave me the explanations I needed to make sense of my own ham radio experiments.

A radio signal is an electromagnetic wave whose behavior is explained in part by its frequency – the number of variations – thousands, millions, or even billions – per second. Waves of different frequencies behave differently. I was beginning to understand these behaviors.

But I had no idea where radio – and the bad boys – would take me.

CHAPTER 2

BROADCASTING TO ESKIMOS

NOVEMBER 1973

The Arctic wind stung my face. The feeling was more pain than cold. I steered the snowmobile into the early morning darkness and turned onto snow-covered tundra, pointing the machine in the direction of the radio station's transmitter building. There I would check that things were operating properly – part of my work as manager and engineer.

My radio skills and a craving for adventure had taken me to the last frontier – Alaska.

The snow seemed brushed with a broom, piling in an eddy behind every object that disturbed the wind's path. Blowing snow hid the transmitter building that was perched near an invisible coastline. Snow-blanketed ice and frozen tundra blended together, obscuring the boundary between land and sea. The Arctic in winter is a moonscape on earth.

Public radio station KOTZ served Eskimo communities spread along the rivers and coasts of remote northwest Alaska.

Its studios were tucked into the schoolhouse in the village of Kotzebue, a small cluster of buildings huddled on a gravel spit jutting into the Chukchi Sea. For centuries Kotzebue's people had subsisted on the beluga whale, bearded seal and ring seal they took from the sea and the caribou and moose they hunted on land.

The village was 30 miles north of the Arctic Circle. Its small wood frame houses had replaced the traditional sod huts that once squatted on the gravel spit. The transmitter building was less than a mile from the houses, but there was no winter road.

I wore multiple layers to blunt the wind's chill. The layers were topped by insulated snow pants, Eskimo-made boots of caribou hide, down-filled mittens, and a parka with a hood trimmed in wolf fur. A knitted balaclava enveloping my head and face had a narrow eye slit that was beneath my goggles. I was completely covered. Exposed flesh would have frozen in less than a minute.

In the Arctic interior, winter temperatures can plunge to -60 F, and winds can cause chill temperatures as low as -100 F. But in Kotzebue, with its milder coastal weather, temperatures rarely drop below -30 F. The wind compensates though. Wind speeds to 30 or 40 miles per hour drive chill temperatures to -100 F, equaling those found in the interior.

I opened the door to the small transmitter building, and warm, soothing air rushed over me. I shook the snow from my parka and, once inside, stood under the electric space heater. The radio transmitter should have made enough heat to warm such a small building, but in the Arctic an extra boost from the heater was needed.

As my body thawed, I shed my parka and mittens and began to work. The building's bare white walls surrounded my toolbox and four racks of equipment. In three of the racks, unadorned, gray door panels covered the innards of the 5000-watt transmitter that pushed the KOTZ signal hundreds of miles across northwest Alaska. Only a row of meters was visible across the transmitter's upper edge. I opened the gray panels, exposing a matrix of knobs and smaller meters that I would use to make adjustments. The fourth rack of equipment held the meters and dials that were part of the studio-to-transmitter radio link and remote control system that allowed the transmitter to be controlled from our studio back in the village. The equipment radiated heat and the smell of dust.

My work was minor, and I would soon be back in the biting wind. I needed only to read a few meters and turn a few knobs to make adjustments. But, when more serious technical problems arose, I would peel off layers of warm clothes and settle in to do repairs. I sometimes plugged in a soldering iron and used it to install new parts, with the smell of the solder's rosin and a curl of smoke drifting toward my face.

Unlike other public radio stations of the day – mostly FM stations – KOTZ was on the usually commercial AM band because of its better coverage. We were only a 5000-watt station, but our listening area was huge, covering more than 100,000 square miles. The AM frequency allowed our signal to be picked up in villages spread all across northwest Alaska.

AM stations use a range of radio frequencies – technically called *medium frequencies* – that were familiar to our listeners as the numbers between 540 and 1600 on the radio dial. These numbers represented frequencies from 540 to 1600 thousand variations, or cycles, per second – 540 to 1600 kilohertz – far fewer than the millions of cycles per second of FM radio stations.

KOTZ's medium frequency waves behaved differently from the high frequency ham radio signals I had sent out from the K2ZMO ham shack. And their behavior was even more different than the very high frequency – VHF – and ultra high frequency – UHF – signals I would later use in building rural telephone systems and the first Wi-Fi network. Each band – medium frequency, high frequency, VHF, and UHF – has unique properties, and medium frequencies are well-suited to covering a large listening area like the one that was served by KOTZ.

On the AM band there are two important ways that radio signals can travel from a transmitting antenna to a receiving antenna, by *ground wave* and by *sky wave*. Ground wave signals provide steady, 24-hour reception in a station's primary coverage area, reaching listeners up to a one or two hundred miles away. Day and night, ground wave signals are always there.

Sky wave signals are the ones that *skip* at night, bouncing off the ionosphere. It's the same effect that once allowed my ham radio signals to reach friends around the world. An AM station's signals usually skip only at night, allowing listeners to hear faraway AM stations that are sometimes strong enough to overpower even local ones.

But a 5000-watt AM station's ground wave signals would not normally have covered an area as large as ours. Our most distant listeners strung up wire antennas to improve reception. They really wanted to hear KOTZ. There was less radio noise and interference in Alaska's villages than its cities, and this too helped the villagers pull in our signal. Even in the village of Kobuk, about 250 miles from Kotzebue, reception was good.

Whether they listened on a portable, battery-operated radio while fishing for summer salmon, or huddled near a cabin's wood stove late on a winter night, our regular listeners received steady, reliable signals because of the station's ground wave signal. But KOTZ's sky wave signal reached listeners spread over an even bigger area. Like ham radio transmissions, KOTZ's nighttime signals could skip to Russia, Japan and other countries across Asia. Letters and postcards often arrived from listeners in these faraway places.

Ground waves travel along the earth's surface, conducted in the same way that an electrical current is carried by a wire. How well this works depends on how well the ground conducts the waves, and that depends on the radio station's electrical connection to the ground. A system of ground wires makes the critical connection. We had laid lots of long copper ground wires over the surface of the tundra at our transmitter site. I put up signs to warn passing snowmobilers, who might run afoul of the ground wires in places where wind had blown the tundra bare. The copper wires were a hazard, but they were necessary for the radio station's signal to reach the villages. We needed them to serve our listeners.

They were *radial* wires, extending outward in all directions from the station's antenna like the spokes of a wheel. At most AM stations, the wires were buried, but Kotzebue's ground conditions made that impossible.

Just below the tundra vegetation surrounding our transmitter, lay *permafrost*, the permanently frozen ground found in western, northern, and interior Alaska. Soils with permafrost can cause havoc with roads and buildings that have conventional foundations. They can also cause serious problems for electrical and communications systems. And, where the rock hard permafrost begins within a few inches of the ground surface, as it does in Kotzebue, digging is almost impossible. That's why we put our radial wires above ground.

Even with signs warning about the wires, passing snowmobiles and all-terrain vehicles sometimes broke them. Repairing ground wires on the tundra was one of my tasks as the station's manager-engineer.

After making adjustments to the transmitter, I locked the small building, remounted the snowmobile, and pulled its starter cord. With the engine running, I turned toward the KOTZ studio to get ready for morning sign-on. It was 5:30 AM, still long before the winter sunrise. The village's twisting streets were quiet. But, after the 6 AM sign-on, Kotzebue would begin to come alive.

Leaving the cold again, I entered the warmth of Kotzebue's only school, loping down the stairs to our basement studios to begin preparations for the broadcast day. A year earlier, because no other space was available, the studios and office had been built in the basement of the school. Yet, even with heating pipes criss-crossing overhead, the station was functional and comfortable.

We had two studios, an office, and a central area used to receive visitors and socialize. There were no outside windows, but interior windows were many. It was, after all, a radio station. Later in the day, these interior windows would reveal frenetic activity in the studios. A visitor's eyes would be drawn to announcers spinning records, twirling dials, and doing on-air interviews.

KOTZ was one of a new breed of radio stations that had started to pop up in America in the late 1960s. Public radio stations were meant to be alternatives to the commercial stations that provided service to most of the United States. The commercial stations didn't often provide programming that appealed to certain small audiences. People who liked classical music, folk music and jazz were examples. The thinking was that commercial stations had no economic incentives to serve such small audiences and that, as long as they were supported by advertising, things wouldn't change. And so, public radio was born.

Many of the new public radio stations began broadcasting on the FM band – from 88 to 108 on the dial – where licenses were easier to get than on the more crowded AM band. And FM's high-quality stereo transmission was ideal for the classical and jazz music sought by public radio listeners. Public radio wanted to provide a service that appealed to a sophisticated audience – one whose numbers were small.

But we were different. KOTZ, officially a public radio station with no advertising support, didn't fit the pattern. We served all the listeners in northwest Alaska, and we sounded more like a commercial station. In fact, a few of the nation's mainstream public broadcasters were horrified to hear that there was a public radio station in Alaska that was playing top 40 and country music.

But our listeners were happy, and I was happy. We were meeting an important need. There were no commercial stations serving northwest Alaska, and there was no prospect for advertiser-supported radio in such a remote part of the state. The same was true in other parts of Alaska. Other public radio stations, with music like ours, began to appear across remote parts of the state in the 1970s.

My first task at the station was to make coffee. Next was our teletype machine, the electromechanical contraption that connected us to the Associated Press wire service, the source of our state, national and international news. But the teletype machine was a technological relic. Sitting on the floor in the corner of our reception area and standing waist high, the machine clackety-clacked as it typed out the day's news relayed from distant cities of the nation and the world. The wire service "copy" would later be ready for our announcers to "rip and read" the day's news to KOTZ listeners.

Teletype was an early form of digital technology. Something like Morse code telegraphy, with its *dits* and *dahs*, teletype messages were sent as a series of current pulses, short periods of current flow – called "marks" – followed by similarly short periods of no current flow – called "spaces." Today's computer geeks might call them "ones" and "zeros." Just as a modern computer interprets a certain pattern of ones and zeros as an "A," a teletype machine interpreted a particular pattern of marks and spaces as an "A." The teletype patterns were different from the patterns used by today's computers, but the idea was the same.

Translating the current pulses into written text needed a jumble of mechanical and electromechanical parts. Together they clicked and clacked in a recognizable cacophony that told me the teletype was working properly. The machine's complexity included relays, gears, arms and levers, all needing my tender care to keep them adjusted, oiled and working properly.

But my early-morning task was simpler – just to change the machine's paper and ribbon. We kept rolls of teletype paper to satisfy the machine's voracious appetite. And the ink-laden ribbon – similar to the typewriter ribbon of the day – had to be changed often. In the 1970s and before, these were familiar chores to radio station engineers and announcers everywhere.

With the coffee pot perking and the teletype machine resupplied, it was time for sign-on. I pushed a button, and the remote-control equipment brought to life our transmitter – in its small building on the snow-swept tundra. Pushing another button started our recorded sign-on and the broadcast day.

Sign-on began with the Alaska Flag Song. It was Alaska's "national anthem."

Eight stars of gold on a field of blue —
Alaska's flag. May it mean to you
The blue of the sea, the evening sky,
The mountain lakes, and the flow'rs nearby;
The gold of the early sourdough dreams,
The precious gold of the hills and streams;
The brilliant stars in the northern sky,
The "Bear" — the "Dipper" — and, shining high,
The great North Star with its steady light,
Over land and sea a beacon bright.
Alaska's flag — to Alaskans dear,
The simple flag of a last frontier.

The words and music were well-known to every school-child and nearly every adult Alaskan from Arctic villages to the far away cities of Anchorage and Fairbanks. The song and its namesake flag bound Alaska's people together. Most Alaskans knew the words of their state song by memory, and they revered the flag's "eight stars of gold on a field of blue." The eight stars are the most prominent ones in the northern sky, the Big Dipper – also called the Great Bear – and the North Star.

Next came the pre-recorded voice of young Joe Hill, one of our staff announcers. He made the familiar sign-on announcement: "Welcome to another day of broadcasting here at KOTZ...owned and operated by Kotzebue Broadcasting Incorporated, KOTZ uses a frequency of 720 kilohertz at a power of 5000 watts, as authorized by the Federal Communications Commission..."

With the sign-on complete, it was time for the "Alex in the Morning" radio show. I was, by this time, wide awake, and my job was to help listeners start the day. The show began with upbeat banjo theme music, followed by a voice-over: "Good morning northwest Alaska! Hello Kiana! How ya doin', Shungnak," I said, naming two of the villages in the listening area. The morning news, weather forecasts, and a blend of top 40 and country music weren't far behind.

Soon other staff members began to appear at the station. First, secretary-receptionist-bookkeeper Phyllis Harris arrived to open the office and make the second pot of coffee. Then young Nellie Ward arrived to prepare for her on-air shift, which came right after mine.

Weather forecasts were a critical part of our service. Several times a day, we relayed detailed forecasts from the weatherman on duty at the small Kotzebue office of the National Weather Service. The region's people listened intently, especially if they were planning to travel.

People journeyed from village to village by snowmobile, making it important for them to know about approaching storms. High winds and blowing snow could hide trails and make travel treacherous. Some trails crossed sea ice, adding another element of danger. Ironically, travel was most dangerous when the weather was warm. High temperatures signaled storms approaching from the south, bringing high winds and blowing snow, sometimes causing the sea ice to break up.

But, before the National Weather Service brought its modern weather technology to Kotzebue, an old-timer named Ed Ward discovered a key to accurate weather forecasting. Ed had lived in Kotzebue a long time, working as a Federal Aviation Administration flight service specialist, a weather forecaster, and – most important to me – a radio telegraph operator and radio ham with call sign KL7BVX. Local bush pilots depended on Ed for up-to-date weather information, often stopping at his house for a cup of coffee and a report on the latest conditions.

To improve the accuracy of his weather forecasts, Ed picked up transmissions from our neighbor just to the west – Russia. He expanded his Morse code skills, mastering all the letters of the Russian alphabet, which includes many of the same letters as English, with the same pattern of *dits* and *dahs*. But Russian also has other letters and *dit-dah* patterns not found in English. With the extra letters, Ed could receive weather stations transmitting from eastern Russia, just to the west of Kotzebue. The Russian stations' signals were loud and clear, easily readable by the experienced Morse operator. And international Morse was alive and well in northwest Alaska.

At 720 on the AM dial, KOTZ's signal carried the voices of young announcers Joe Hill, Carolyn Smith, and Ed Ward's kids, Nellie and Delbert – both of them worked at the station. All recent graduates of Kotzebue's high school, the four announcers spread energy to the distant villages. Their voices projected pride in the new station and the service they were providing.

Nellie happily sent an invitation to one of the villages: "My cousin in Noorvik wants to let all the village people know that she made lots of *agutaq* (Eskimo ice cream), and she wants to share it tonight. Come on over if you want some."

Carolyn had an important announcement for another village: "Health aide Lucy up in Kiana has flu shots available for all the kids, but *umuk* the babies (carry them on your back inside a warm parka) because it's cold out there!"

At the age of 19, Nellie was a memorable character. Full of life, her broad cheeks and straight, black hair betrayed her Eskimo ancestry. With an Eskimo mother and a white father, Nellie had learned to live in two worlds. Equally comfortable wearing a *kuspuk* (Eskimo woman's dress) and eating *muktuk* (whale blubber) at a village celebration, or interviewing visiting state officials about government policy, she was irreverent in a way that made me laugh. ("That guy from Anchorage was really full of ...") Her charm and dry humor flowed through the microphone and spread across northwest Alaska. She was a star. Nellie later became one of Alaska's leading journalists.

Joe Hill had spirit, too. A year younger than Nellie, Joe was of medium height and powerfully built. He was interested in radio technology and he sometimes helped with my engineering tasks, asking lots of questions and eager to pitch in. Joe lived with his mother and stepfather, who ran one of the general stores in town. He had graduated from high school just a few months before my arrival in Kotzebue, and, like Nellie, he had a sharp, dry sense of humor that was evident as we traded good-natured jabs. ("Is that really how you do things in the big city?")

Joe quickly saw inconsistencies in the pronouncements of older people – including me.

Carolyn and Delbert were good announcers and faithful employees, but they didn't push back like Nellie and Joe. I was glad to have them because they lent steadiness to the station. So did secretary-receptionist-bookkeeper Phyllis, who was older than the rest of us. She was the den mother. And Meg, the beautiful young nurse who would soon be my bride, often visited the station. We were a family.

The heart of KOTZ's programming was the music, a blend of top 40 and country and western. Among our top tunes in 1973, our first year on the air, were:

Top of the World – The Carpenters

Bad, Bad Leroy Brown – Jim Croce

Delta Dawn – Helen Reddy

Killing Me Softly with His Song – Roberta Flack

Tie a Yellow Ribbon Round the Old Oak Tree – Tony Orlando and Dawn

You're So Vain – Carly Simon

Kodachrome – Paul Simon

Satin Sheets – Jeanne Pruett

Why Me? – Kris Kristofferson

At night, when listeners moved close to the radio, we had "Eskimo Stories." Kotzebue's elders took turns telling traditional stories in the Inupiaq Eskimo language. Old-timers –

some could remember the introduction of reindeer to western Alaska in the early twentieth century – were regulars. Through their stories, they helped to preserve the vanishing oral legends of the Inupiaq of northwest Alaska. After they passed on, other elders followed, and the program continues today, renamed "Inupiaq Stories."

We also had "Old Time Radio." An enterprising company was syndicating radio shows of the 1930s, 1940s and 1950s, and they sent the shows on tape to radio stations across America. Every few weeks we received a box of programs like The Lone Ranger, The Shadow, or The Green Hornet. But in northwest Alaska Old Time Radio wasn't really old.

KOTZ listeners could hear the "William Tell Overture," theme music for The Lone Ranger – or The Shadow's opening line, "Who knows what evil lurks in the hearts of men?" – followed by a 30-minute drama from the past. I had listened to these same shows – exactly the same stories – while adjusting the cat's whisker on my crystal set many years earlier. But there were only a few others in Kotzebue who had already listened to the programs. Most were hearing them for the first time. Old Time Radio was all new to most of our audience, and they loved it.

The telephone never stopped ringing at KOTZ. Listeners called to send the messages we named "Tundra Telegrams," personal messages directed to family and friends in other villages.

Delbert intoned, "Uncle Willie, my plane arrives in your village at 6:30. Please meet me at the runway," followed by "Sister Rachel, Granny left her medicine on the dresser. Send it as soon as possible."

Tundra Telegrams informed extended families in the outlying villages of the passing of a loved one. Or they told the happy news of the birth of a new family member. In harsh weather the messages were a rallying call for volunteers to search for a missing snowmobiler or dog musher. If a family home was destroyed by fire, they called for donations. The messages were an essential service. KOTZ was often the only available means of communication.

And other radio stations in western Alaska sent similar messages. Each station had its own name for the service – our Tundra Telegrams were Village Hotlines or Caribou Clatters elsewhere. It was technically illegal for a broadcast station to send a personal message, but the FCC recognized the importance of the services and focused its attention elsewhere.

Services like Tundra Telegrams highlighted the inadequate communication systems in villages across Alaska. In 1970, when I first arrived in the young state, there were, with few exceptions, no radio stations and no telephones in rural Alaska. Within a few years, KOTZ and other public radio stations filled one gap, but there was still no telephone service in the villages.

A typical village might have one or two shortwave radio sets for communication to the outside world. But shortwave signals, like the ones I had sent from my teenage ham radio transmitter, didn't provide reliable communication in the north. As a ham operator, I had been excited each day to try to communicate with a distant friend. But the villages needed reliability – not excitement.

One problem was the *aurora borealis* – the northern lights – an electromagnetic storm in the ionosphere. This is the same atmospheric layer – 60 to 210 miles high – that reflects ham radio and other high-frequency signals, allowing long-distance communication. The northern lights are a beautiful sight, with wispy curtains of blue, green, yellow and pink, waving across the northern sky. But a particularly colorful display could kill high-frequency communication for a week or two. That meant no two-way communication in northern Alaska.

((•))

Alaska's hundreds of small villages needed telephone service, and Alaska's long-distance telephone company – RCA Alaska Communications, Inc. – created the "bush telephone program" to provide the service. I led the bush telephone field crew. Our job was to put a single VHF radio telephone in each village. The VHF signals followed direct paths from transmitting to receiving antennas, avoiding the *aurora borealis.*

With frequencies from 30 to 300 million cycles per second – 30 to 300 Megahertz – VHF signals have different characteristics than the ones I used at KOTZ and as a radio ham. The higher frequencies approached those that I would later use in building the first Wi-Fi network.

The equipment we used was really meant for the trunk of a car. It was designed to provide a kind of mobile telephone service that preceded today's cell phones. But we adapted the vehicle equipment to be installed at a central place in a village, a place like the general store or medical clinic. Phone calls were transmitted through a *yagi* antenna – it looked like a TV antenna standing on edge – to a base station we erected in a bigger town that already had phone service. From there the call went through the telephone network like any other call.

VHF signals are *space wave* signals that travel in a line between transmitting and receiving antennas. Since they don't reach the ionosphere, where the *aurora borealis* causes problems, reception is more consistent than with shortwave radio. This made the VHF system more reliable. For the first time, people had a good way to call the doctor or order badly needed snowmobile parts. Or just call Johnny, away at school for the year.

Our little band of equipment-installers was four strong. Two technicians, a pilot and I crisscrossed Alaska, installing radio phones as we went. The technicians came out from Anchorage to work for a week or two each time. Then they were relieved by replacement technicians.

Pilot John Lee flew our Cessna 185 on pontoons – we called them "floats" – in the summer and on skis in the winter. John was also a civil engineer, and he helped me to design antenna installations and choose the best location in each village to install our equipment. Technicians rotated in and out of our team, but John and I were always on the job. We became close friends. Before technicians could install equipment, we made rounds of the villages together on special trips needed to decide where equipment would be installed.

A typical "site survey" trip began when John and I met at Lake Hood, Anchorage's airport for sea planes – float planes. The unusual air strip is a canal cut between two lakes that lie beside Anchorage's International Airport, home to big jets flying to the other 49 states, Europe and Asia. On instructions from the tall tower at Anchorage International, a small float plane can taxi from its dock, line up on the canal runway, and fly off to remote Alaska, later landing on any convenient river, lake, or bay.

John always had the plane ready when I arrived. Painted red and white, the Cessna 185 bobbed in the water on its silver-colored floats. I pulled on my rubber hip boots – extra long to accommodate my 6'4" frame – untied the plane from the dock, and climbed into the copilot's seat to John's right. Also wearing hip boots, John started the engine, and our Cessna began to taxi.

"Anchorage tower, Cessna seven-zero-zero-two-two taxiing for departure Lake Hood," John said into the microphone. After receiving clearance from the tower, he pushed the 185's throttle forward, the engine roared, and we rushed down the watery runway. Our floats broke free of the water, and we turned west toward bush Alaska.

After landing at a village and tying up the Cessna, I looked for the village chief, mayor, or another responsible official. John kidded me, saying that this was just the "public relations" work usually disparaged by engineers. But communicating with local officials could be as important as our technical work. I needed to explain the bush telephone program to the people in charge and reach an agreement on where the equipment would be located.

And my work with village leaders was important in other ways. Though the bush telephone project was at first technically successful, it failed in some villages for a non-technical reason. All calls were long-distance calls and came with a charge. But sometimes telephones were disconnected because of villages' failure to pay their monthly bills. With only one telephone in a village, a method of accounting was needed to charge each long-distance call to the villager who made the call. The company had not anticipated this need.

To set up a village billing procedure, I made a telephone log and asked each village chief to select a telephone attendant to coordinate the use of the telephone. Then I taught the attendant how to account for the long-distance calls. At the end of each call, the attendant would ask the long-distance operator for "time and charges" – the duration and cost of the call. The attendant entered this information into the log, along with the name of the person who had placed the call. The caller could pay right away or settle up at the end of the month. With this system in place, there were fewer unpaid bills – and fewer disconnected phones.

On site survey trips, John and I chose the best location for our equipment and antennas. After seeking the advice of local leaders, we searched for places that had a direct radio path to our base station. We knew that, unlike lower frequency signals characterized by ground wave and sky wave behavior, VHF signals traveled along a direct path from transmitting to receiving antenna.

But the bad boys of radio were lurking. One of them was *shadowing*. A physical obstruction could block – or partially block – a signal's direct path. A bush telephone signal might penetrate the walls of a wood frame house, but the shadowing effect could make it weaker inside the house. That's why we used antennas mounted outside the buildings. And terrain features, like nearby hills and mountains, could also cause shadowing. A VHF signal had little chance of penetrating a mountain. We always used outdoor antennas mounted high enough to avoid the shadowing caused by obstructions like hills, mountains and nearby buildings.

Another problem was *reflection*. When a transmitted VHF signal encounters a smooth, flat surface, maybe a metal surface, it can be reflected, just as a light ray is reflected from a mirror. We usually considered VHF reflections to be a bad thing. Reflections off the ground and the surfaces of lakes and rivers could cause problems when, at the receiver, the reflected signal recombined with the direct signal.

In the villages, our high antennas helped to avoid reflections. We mounted the antennas high enough to assure that a direct path from transmitter to receiver was well above the terrain in between. But there was one village where I used reflection to *improve* reception. More on that coming up.

A third problem was *refraction*. VHF signals don't usually travel in straight lines – instead, they follow curved paths. They are *refracted*. Radio waves – and light waves – normally curve toward the ground as they move through the atmosphere.

I first noticed light's refraction in Kotzebue. Summers and winters brought extremes of light – continuous daylight in summer and only a little daylight in winter. Hidden in these extremes was the phenomenon of refraction.

Northern Alaska really is the "land of the midnight sun." And Kotzebue's ever-present summer sunlight is easy to like. On June 21 – the longest day – the sun makes a big circle in the sky, never dropping below the horizon. On this day, and during the 20 days spanning the period before and after June 21, there are 24 hours of daylight. But the situation flips in winter, when it's dark almost all of the time. On the shortest day, December 21, only one and a half hours separate sunrise from sunset. The sun appears briefly above the southern horizon but soon sets again. The joy of the summertime sun is offset by the long darkness of winter.

I was at first puzzled that there was no parity between the number of daylight hours on the longest day and the number of dark hours on the shortest day. Thinking of the geometry of the Earth on its tilted axis and its orbit around the sun, I imagined that the symmetry of the situation should make the number of dark hours in winter and light hours in summer exactly the same. Straight-line geometry told me that, on the Arctic Circle, the sun should dip momentarily below the horizon on June 21 and momentarily peek above it on December 21. But actually the Arctic Circle receives more daylight than this on both the longest and shortest days. How can this be?

Refraction is the answer. We often think of light as traveling in straight lines called rays. But refraction causes light rays to deviate from straight lines, curving toward the earth's surface as they travel through the atmosphere. This allows the Arctic Circle to receive more daylight, in both winter and summer, than if light traveled in straight lines. Light and radio signals are both electromagnetic waves, and VHF space waves curve toward the ground just like rays of light. That's why refraction is a factor in designing VHF radio systems.

And our bush telephone designs could have considered *scattering* and *diffraction*. *Scattering* occurs when a radio signal strikes an object with a rough surface – maybe a tree. In a way, the signal is reflected, but the tree's roughness sends radio energy in more than one direction. A rough surface has an abundance of small surfaces, each like a small mirror reflecting the signal. Many mirrors send signals in many directions.

Diffraction lets a radio signal bend around a building's corner or any other well-defined edge. Even when a straight line from transmitter to receiver is blocked by one or more walls of a building, diffraction may still allow a radio wave arriving at an edge to make a turn and continue in the direction of the receiver.

Ideally, John and I should have considered all of these effects in designing radio phone installations. But we really needed only to be sure that our antennas were high enough to provide an unobstructed direct path to the radio unit at the other end of the link – with lots of extra room to clear obstacles. If we did this, we had little need to worry.

Yet the five bad boys of radio – shadowing, reflection, refraction, scattering and diffraction – were the same ones that would plague me years later, when I built the first Wi-Fi network. Then I wouldn't be able to use high antennas. I would have to take the bad boys head-on.

With our high antennas, we didn't usually worry about reflection. But our installation at Alaska's most remote village was a different story. The village of Ignaluk is on Alaska's Little Diomede Island, near the middle of the Bering Strait, which separates the Alaska mainland from Russia. The island is usually surrounded by sea ice frozen solid. Winds and sea currents push the ice into huge stacks called "pressure ridges," some of them 20 to 30 feet high.

It was early April, and the Cessna 185's summertime floats had been replaced by skis. Landing the ski-equipped airplane meant finding a place on the sea ice that was free of pressure ridges. Ignaluk had no airstrip. The village could be reached by boat during the few months of summer when the Bering Strait was liquid, but winter access by air was possible only during March and April, when a few patches of sea ice near the island were smooth enough to land a small plane on skis.

From the air Little Diomede Island looks like a mesa sitting, not in the desert, but in the Bering Strait. It's a huge rock with a high, flat top and steep sides plunging to the sea ice below.

Ignaluk is tucked at the base of the cliffs on the west side of the island. The village faces Big Diomede Island, Little

Diomede's larger, Russian brother, a few miles away, across the International Date Line and the United States-Russia boundary. The bigger island was uninhabited except for a military observation post, but villagers told me they sometimes trained a telescope on the Russian outpost and saw a soldier with a telescope looking back at them.

John lined up on a smooth patch and landed on the sea ice. We came to a stop near the village. He kept the plane's engine idling, and, with the nose pointed into the wind, the airspeed indicator read 50 knots – just over 50 miles an hour – indicating the speed of the wind, not the plane. Thinking the wind would soon die down, we unloaded our equipment and prepared to install the radio phone and its antenna.

There was no safe place to tie up the Cessna. John flew back to Nome, leaving one technician and me to work on the island. First we worked indoors, installing the radio, and the next day we assembled the antenna and its mast, laying it on the ground ready to be hoisted into the air when the wind died down.

A few days passed, and, with wind still at 40 miles an hour, one villager told me "wind never stop." There seemed to be no choice but to try to raise the antenna in the high wind. This would have been impossible for our little two-man crew, but the villagers, at ease working in the harsh weather conditions, volunteered to help. They thought nothing of scrambling across icy rooftops in a gale. Few words were spoken, but the villagers quickly raised the antenna to an upright position.

A bystander would have said we pointed the big antenna the wrong way. The Alaska mainland was to the east, and the antenna faced west. But, if we had pointed the antenna toward the mainland, its signals would have run into the high cliffs behind the village with no hope of finding their target.

We couldn't install the antenna in a place with a clear view to the east, but we did the next best thing. We aimed the antenna directly at the cliffs of Russia's Big Diomede, where signals could be reflected back to the Alaska mainland. With the villagers slowly turning the antenna, first right and then left, I watched the indoor equipment's signal strength meter until we found an antenna position where the signals perfectly bounced off the big island's cliffs and found their way to the base station on the mainland. Reception was strong and reliable.

I made a few test phone calls before calling John for a ride home. This time reflection wasn't the problem. It was the solution.

CHAPTER 3

TAKING THE INTERNET WIRELESS

NOVEMBER 1993

We hunched over our sandwiches in the crowded university cafeteria. The room around us hummed with restless conversation. Students bolted down their food and compared homework solutions before rushing off to classes. And the two of us struggled with a problem – a problem we had made ourselves.

Marvin Sirbu and I, both professors at Carnegie Mellon University, had worked together to create a new institute within the university. But the fledgling institute needed our help. It didn't quite fit the Carnegie Mellon mold, and its future might be in doubt. We were both a little worried.

The cafeteria might have been a place for socializing and relaxing, but it was instead an extension of nearby classrooms and laboratories. Class session blended into lunch hour, and lunch hour blended into laboratory experiment, with professors and students moving at a frenetic pace. There was no break in the action.

The hyperactive cafeteria chatter reflected the rhythm of the institution. Carnegie Mellon University, once just a good engineering school, was now world famous, propelled by the energy that pervaded the place. Professors and students seemed driven by an unseen force pushing them to work harder.

For years the university had been an underdog in the world of engineering schools – not a top tier institution like MIT, Caltech or Stanford. Now we had finally joined these prominent research universities. But no one in the cafeteria seemed to realize that we'd made it. We weren't relaxing.

Carnegie Mellon's rise to the top tier had been propelled by technology innovation and by a unique ability to do interdisciplinary work. And, since 1977, when I had first enrolled as a graduate student, I'd been impressed by the way that professors and students easily crossed academic barriers to solve any problem at hand. Other universities gave lip service to interdisciplinary work but with only superficial commitment. At Carnegie Mellon it was just the way people worked.

Technology innovation and interdisciplinary work were our strengths. And both were embodied in the work of our friend and role model, Professor Herbert Simon. I had talked with Herb just a few hours earlier.

Arriving at the university earlier that day, I walked across the campus to my office. The crisp November air was still, and the morning sun, shining through a cloudless sky, had begun to warm the campus. The foliage, green just a few weeks before, showed its November colors of red and yellow. There was the look of a traditional university campus – matching yellow brick buildings with the arched windows and grand entrances of the *beaux arts* architectural style that was popular when they were built in the early twentieth century. But the tranquility was broken when, between classes, students rushed along walkways crisscrossing the big L-shaped grassy area with its two wide segments called "the Cut" and "the Mall."

Swept along by a stream of students, I saw Herb Simon ahead. We were both headed to our offices in the ivy-covered Baker Hall. I slowed to match the pace of the older man, and the eternally late students rushed around us.

As I caught up with Herb, I thought of a recent campus display showing a part of Carnegie Mellon's history. It held a collection of early computer equipment – and photographs of Herb Simon and his colleague Allan Newell. The equipment was an artifact of the birth of a new science, and the photos showed the two men working on the equipment. When the photos were made, the equipment was new and the men were young, but the same equipment, displayed in real life beside the photos, with electrical switches and display lights from another era, had grown old.

Herb and Allan were the founders of "artificial intelligence," the use of computers to emulate human intelligence. The discipline was a combination of psychology and computer science – Herb was the psychologist and Allan the computer scientist. Their creation of artificial intelligence was an example of work combining more than one academic field, the kind of interdisciplinary work that was commonplace at Carnegie Mellon.

Herb was in his seventies, but he was still an active and energetic researcher, working every day at the university. I was preoccupied with my computer work but suspected that Herb was thinking about bigger things. He didn't describe them to me. Instead, he asked about my work.

Slowly, we walked and talked. Herb asked, "What's the position of Apple Computer in the personal computer market?" The question surprised me, and I thought he might have asked it to be polite. Then I realized his curiosity was genuine. He seemed surprised when I told him Apple Computer had only a small fraction of the personal computer market. I explained that university students and professors used more Apples than those in the outside world. Herb was interested in what I had to say.

He was "Mr. Carnegie Mellon." Herb taught at the university for 52 years, beginning soon after he earned a political science Ph.D. at the University of Chicago. During the long span of his interdisciplinary career, his study of human behavior led him to venture without hesitation into other fields – fields like computer science and economics. Herb won the highest honors in economics, psychology, and computer science, among other disciplines. In 1978, as a result of his ground-breaking work on the interplay between human behavior and economics, he won the Nobel Prize in Economics.

Herb appreciated the honors, but his work was more important. He was a full-time researcher nearly until the time of his death, at age 84, in 2001.

Years after Herb Simon and Allan Newell combined computer science and psychology to create artificial intelligence, Carnegie Mellon achieved another big breakthrough, transforming the way computing was done. It happened in the early 1980s, and it was called the Andrew project. Before Andrew, computing was done one way. After Andrew, everything changed.

In computing's earlier days, computers were big mainframes – equipment racks filled with electronics, dials, switches and lights – installed in air-conditioned rooms. These huge machines could be used in either of two ways. First, you could write a program, use a special "key punch" machine to code your program as holes in special IBM cards, and then take the deck of cards to a computer center. There operators worked, priest-like, to serve the mainframe, with its array of flashing lights whose meaning was known only to the highest of the priests.

An operator stacked your card deck into a card reader, a smaller machine that devoured the stack from the bottom as the cards slipped gradually into its maw. The program punched into the cards was processed by the mainframe, and its output was returned to you on strangely wide paper with holes on both sides – holes used by sprockets to move the paper through a big printer.

You could also interact with the mainframe by using a "time-sharing terminal" connected by wires to the computer. The terminal, with a keyboard and display, could be feet or miles from the mainframe, in another part of the building or another part of the city.

In the 1970s I had used a time-sharing terminal to send e-mail messages, then a little known way to communicate. My time-sharing terminal was the most primitive kind available – a teletype machine. Like the Associated Press teletype in the KOTZ studio, this one went clickety-clack, its gears, relays, and levers in motion as I typed on the keyboard. The clicking and clacking ran in fits and starts, each sound the result of my pressing down hard on a key to enter a single character.

But the sounds became more regular when the teletype machine was printing a message from the mainframe. Then it was more like the rhythm of the old KOTZ teletype. Numerals and letters of the alphabet came across the wire with clocklike regularity because there was a machine – not a human – at the other end.

But a new approach to computing began in 1981, when Carnegie Mellon partnered with IBM to create the Andrew project, ultimately connecting thousands of computers across the campus to central servers and distributed printers through a new high speed computer network. It was a historic first. The Andrew project used what we now call "client-server computing," unique then but now commonplace. The new network was named "Andrew," after Andrew Carnegie and Andrew Mellon, the Carnegie and the Mellon of the university's name. Marvin and I didn't know it as we talked in the cafeteria, but the Andrew name would be part of the solution to our problem.

Talking over the clatter of plates and silverware, Marvin and I struggled with the problem. We had worked together a few years earlier to create the new institute at Carnegie Mellon, and the institute now suffered from a big gap in its activities. We needed to fill the hole.

Marvin, with his high forehead, curly hair and mustache, fit well at Carnegie Mellon. Quick, energetic and creative, he was in constant motion and usually processing more than one thought at a time. He spoke quickly but concisely, expecting others to follow his rapid fire delivery. Students said that listening to his lectures was like drinking from a fire hose.

Yet, with our different styles, we made a good team. Where Marvin was quick, bright, and creative, I was calm and methodical. When we worked together to start the new institute, Marvin provided intellectual energy, and I provided organization, structure, and follow-through.

It was called the Information Networking Institute. In the late 1980s and early 1990s, a period of accelerating innovation, computer and communication technologies were merging into one, and their importance was rising in the world of business. That's why our institute was interdisciplinary. It offered an educational program to train a new kind of professional – one proficient in both technologies and also knowledgeable about business. The institute's new master's degree program was a direct response to the increasing demand for such graduates. We were educating the new breed of professionals.

But Marvin and I also wanted to engage Carnegie Mellon professors in information networking research. We knew that a Carnegie Mellon institute could not be successful without a strong research program.

We needed to come up with a project encompassing the varying parts of information networking and drawing on the talents of existing Carnegie Mellon professors. We were a small institute, and we couldn't afford to recruit new faculty. Yet the project would need to attract outside money from government and private sources.

Talking over the clatter of plates, I asked "what about wireless?" Marvin stopped chewing. His eyes widened. "Wireless?" he said. His mind was racing. I said I wanted to create a new high-speed wireless network – a network that could be used by wireless researchers to do experiments. I thought that we might someday be able to expand the wireless network to cover the entire campus, allowing it to be used by all of our students, faculty and staff. As we talked, Marvin warmed to the idea.

At the time, "wireless" was a word emerging from the dustbin of American usage. A hundred years earlier it had meant, quite simply, *radio*, an innovation that was then grabbing the world's attention.

Guglielmo Marconi first sent radio signals in 1897 over short distances in England, and, within a few years, he had established commercial radio telegraph service between Clifden, Ireland, at the western edge of Europe, and Glace Bay, Nova Scotia, at the eastern edge of North America, forming the shortest possible radio hop between the two continents. The clicks of Samuel F. B. Morse's land-based telegraph were soon supplemented by the *dits* and *dahs* of radio telegraph for communication with ships at sea. Wireless made seafarers safer and kept them in contact with their home ports. And it made trans-Atlantic news dispatches possible, bringing Europe and North America closer together.

The word "wireless" was later used, especially in Great Britain, to mean a radio receiver, as in "what's on the wireless tonight?" It was a question asked by British listeners waiting for the counterparts of America's Lone Ranger, the Shadow and the Green Hornet to arrive at their wireless sets. And the word persisted much longer in Britain than in the United States. By the time I became a teenage ham radio operator in the 1950s, it was not used on our side of the Atlantic. An American using the term then could expect to receive strange looks.

But "wireless" began to find new life in the 1990s, when it reentered the lexicon with a fresh meaning – the technologies, all using radio, that would allow voice and data communication service to be used anytime and anywhere. Today's technologies like 3G and 4G cellular, Bluetooth, Wi-Fi and WiMAX are all modern examples of wireless.

Marvin and I had heard the buzz about wireless. Cellular telephones were gaining popularity in the United States and elsewhere. And mobile computers were on the horizon. Futurists talked about the day when a computer would be carried in a shirt pocket and remain connected to a computer network, wherever its owner went. It seemed like science fiction, but there was an obvious synergy between mobile computing and wireless communication. When computers became truly mobile, wireless networks might allow seamless, anytime-anywhere connection to the Internet.

Cellular telephones had been invented at Bell Telephone Laboratories in the early 1960s, but it took twenty years for the Federal Communications Commission to finally allow them to be used by the public. After the FCC's decision in the 1980s, new handheld sets appeared on the market. Previously, a mobile phone was a large box riding in the trunk of a car and connected to a telephone handset used by the driver. Even then, using a phone could be a driving hazard.

When Marvin and I talked in the cafeteria, most cellular phones were analog and used the same modulation method as FM radio stations. Widespread use of more advanced digital phones would come later. And using radio to provide computer connections was rare. But these were the connections that really interested me. I wanted to be one of the first to use wireless to interconnect mobile computers. There were companies already offering this kind of service – sort of.

Two of them, RAM Mobile Data and Ardis, provided services to mobile workers like repairmen and delivery men, constantly on the move but still needing to communicate with their home offices. RAM and Ardis services allowed the workers to carry tablet computer-like devices – big and clunky by today's standards – linked to the RAM and Ardis wireless systems. The Otis elevator repair guy, with elevator parts spread on a hallway floor, could use his device to see engineering drawings of the elevator he was repairing, or he could order repair parts while still looking at the innards of the elevator. Package delivery companies like UPS and FedEx also used the new service.

But the speeds of the RAM and Ardis networks were glacial. If you used them for e-mail, you could go out for coffee while waiting for your messages. Their maximum speed was 19.2 kilobits per second – even slower than the dial-up modems of the day.

At the same time, Bell Atlantic Mobile, a Baby Bell cellular company operating in Pennsylvania and other middle Atlantic states, was talking about a new service called CDPD – Cellular Digital Packet Data – that could piggyback on their cellular network. The company wanted to introduce the service in Pittsburgh and then expand it across Pennsylvania and to other states.

Bell Atlantic could offer CDPD wherever cellular service was available because it was just an add-on to the existing cellular network. But, even at its highest speed, CDPD would still be painfully slow. Its maximum speed was only 19.2 kilobits per second, the same as RAM and Ardis. There was hope, though, that cellular networks would eventually offer higher speeds, and I wanted to talk to Bell Atlantic Mobile about CDPD.

But achieving the vision of interconnected mobile computers would require another piece – small and easy-to-carry computers. Computers were already getting smaller – some could even be carried in a briefcase. There was an Apple Macintosh portable – really more "luggable" than portable – but still better than trying to carry around a desktop computer.

Two things were limiting the new, smaller computers – screen and battery technologies. Some companies were trying to make computer screens small enough to be used in a mobile device but also able to show clear images. Battery technology restricted a mobile computer's ability to work while disconnected from an electrical power source. Better batteries would mean more time between plug-ins.

Marvin and I thought that mobile computers would soon be smaller, have better screens, and be able to operate for longer periods without a battery recharge. There was no Blackberry or iPhone on the horizon, but we imagined that someday lots of computing power would be contained in a small package.

It seemed that mobile computers and high-speed wireless networks could be a powerful combination. Just as the Andrew network, with its distributed servers connected to a high-speed network, had revolutionized stationary computing, the combination of mobile computers and high-speed wireless networks could create a new world of untethered, mobile computing. We didn't know how long this would take, but, if we built a high-speed wireless network, it would be ready to work with small mobile computers when they became available.

We also knew that e-mail was beginning to gain public acceptance. The World Wide Web had just made its debut. I imagined e-mail and Web browsers being used with mobile computers. But I also thought that mobile computers could someday do far more than their desktop counterparts. It wasn't just a question of untethering users' desktop machines and continuing with the same activities. There could also be a whole new spectrum of computer applications made possible by mobile computers and wireless networks working together. Today we call them "apps."

A few years later a university photographer captured my vision. His photo showed a young woman sitting on Carnegie

Mellon's grassy Mall and using her laptop computer. Protruding from the top of the laptop was a small antenna. She was surfing the Internet! The picture said it all. It telegraphed the vision of anytime-anywhere Internet access for all users of mobile computers.

But was wireless technology up to the job? Services like RAM Mobile Data, Ardis, and Bell Atlantic Mobile's CDPD were a start. And there was an obscure technology that might be even better. It was called wireless local-area network – wireless LAN – technology, and it was said to operate at much higher speeds than other wireless systems.

Wireless LANs worked within a small area of a building – one or maybe two rooms. No one had built a bigger wireless LAN, one covering an entire building or an entire campus. Yet I wanted to provide wireless coverage to our researchers and later deliver high-speed wireless Internet service to all of the Carnegie Mellon community. Could wireless LAN technology be adapted do the job? I wasn't sure.

I suspected that the bad boys could cause some trouble. Wireless LAN equipment used frequencies in the UHF band, higher frequencies than the Alaska's bush telephone program's VHF band – but with similar problems. I remembered the bush phone program's shadowing, reflection and refraction. They were sure to cause problems in a wireless LAN with a large coverage area.

Marvin and I struggled to hear each other's voices over the clatter of dishes and silverware, but we agreed that a wireless project could build on the strengths of the new institute and on our own interests. Marvin was interested in computer networking, and I had worked with radio, the underlying technology that could support a wireless network. There were other professors at Carnegie Mellon working in areas related to wireless. They would be interested, too. Our enthusiasm rose.

We were on the verge of a decision that this was the right project for the institute when Marvin realized he was late for a meeting. He pushed back his chair, stood up and rushed off.

I lingered at the table, beginning to make a plan. A few of our Carnegie Mellon colleagues were already working on mobile computing, and they would want to join the research initiative. They could use a wireless network to support their work. The combination of wireless and a set of interrelated mobile computing projects could be a powerful mix that would attract other researchers and lead to the new research the institute needed.

The cafeteria was closing. I took my tray to the dishwashers. The room, now empty, had become quiet. I walked back to my office, thinking about wireless and wondering what problems lay ahead.

CHAPTER 4

BAD BOYS LURKING

NOVEMBER 1994

I entered the classroom unnoticed. Chaotic chatter engulfed the room – not the chaos of an undisciplined student group but a jumble of excited conversations like those I would later hear in the conference rooms and laboratories of Silicon Valley startup companies. Students had pulled chairs together into groups. They interrupted and talked over each other. The words that filled the air were not of frat parties or the latest tunes – they were technical words like "bits," bytes," "RAM" and "ROM." The students were feeling the pressure of their project deadline only six weeks away. Their completed project would be small. The challenge would be big.

The chatter stopped abruptly, and the room quieted, when I began to speak. I told the class about the new wireless network. They were intent on my words because they hoped to use it – or some other wireless network – in their work.

It was a project course – students working together to build the ultimate mobile computer, a *wearable* computer. It would be small enough to be worn on your body, hanging on your belt – like the popular "Walkman" radio of the day – and always available to be used. The students' wearable computer would be bigger than today's iPhone or Android phones, but this was 1994.

Both the project course and the wearable computer were the ideas of Professor Dan Siewiorek. Dan thought that wearable computers would someday be worn like jewelry and would have all the capabilities of desktop computers. It was the stuff of science fiction. Think of the Star Trek communicator pinned to the chest of Captain Kirk's uniform, and you will understand Dan's vision. He wanted to build a piece of Star Trek in the twentieth century. The Walkman-size computer was just a step in that direction.

But Dan worked through his students. He inspired them, he motivated them, and he collaborated with them. Each semester, Dan led a project course aimed at building a wearable computer, and each semester's computer was smaller and more capable than the last.

Looking like a stereotypical engineer, with horn-rimmed glasses and a pocket protector full of pens, Dan was soft-spoken and intelligent. Absorbed in his technical work, he chose his words carefully. But he was also likable and gracious – always willing to help students and younger professors. A Carnegie Mellon superstar, Dan had won awards and honors for his work in exotic sounding fields like "formal systems description and evaluation" and "multiprocessor computer architecture." Yet he still seemed like the friendly guy who lived around the corner.

At the time of my lecture, none of Dan's wearable computers were equipped with wireless. But that was about to change

– wearable computers and wireless needed each other. I explained to the class that I wanted the new wireless network to be a campus-wide mobile computing laboratory. It would be a tool to support research, but its creation would itself be a research project – it would be far larger than any wireless local-area network that had come before.

Warming to my subject, I described existing wireless services that could be used to interconnect computers – Ardis, RAM Mobile Data, and CDPD. None of them could provide the high-speed Internet connections that our researchers needed. The existing services' speeds were dismally slow – even slower than dial-up service.

I told the students about the new wireless local area network – wireless LAN – technology. Each wireless LAN had a box, typically called an *access point*, that could be connected to a wired network. And there were other boxes, called *network adapters*, that could be connected to the desktop computers in a room. All of the boxes talked to each other, using radio to form a network linking them together. The data speeds were high – 50 to 100 times faster than the other systems – up to 1 or 2 Megabits per second.

Wireless LAN equipment helped engineers to avoid the problems of using old telephone wiring to make connections between desktop computers. New buildings had modern wiring designed for computers, but the wiring in older buildings was often not up to the job. It would take years to replace all of that old wiring.

A wireless LAN could help – it was easier to install a little equipment than to rewire a building. And the equipment made it easier to rearrange an office space, moving desktop computers with no need to worry about wiring.

But wireless LAN technology seemed to have other possibilities. If a wireless LAN could be adapted for use with a mobile computer, and, if we could figure out how to extend its coverage area beyond a room, we might be able to create a wireless research infrastructure and, at the same time, take a step toward achieving the vision of a wireless campus. They were two big "ifs."

But that day I was an optimist. I told the class that we would someday have a wireless network to provide untethered, high-speed Internet service for all of the university's students, professors and staff, and they would use it anywhere on the campus. Everyone would carry a mobile computer linked to the wireless network. I spoke with confidence, but, in truth, I didn't really know if I could make the dream happen.

At first we would use laptop computers. Still, I thought that we would soon have smaller, mobile computers. And handheld devices couldn't be far away. Dan's wearable computers might someday be commercial products. It was the "anytime-anywhere" vision – anything you could do on the desktop computer in your office could be done on a mobile computer wirelessly connected to the Internet. Dan liked the idea and wanted the class's next wearable computer to use the new wireless network.

After Dan's class, I walked toward my Baker Hall office. The campus looked as it had a year earlier when Marvin and I had hatched the plan to build a wireless network. Red and yellow leaves floated gently to the still green lawns of the Cut and the Mall, and a river of students flowed around me, making their way to classes.

I didn't know if a big – even campus-wide – wireless LAN would work. I focused on a smaller, near-term goal. With the help of the new director of the Information Networking Institute, I submitted a proposal to the National Science Foundation to fund a high-speed wireless network. The network would be a wireless research test bed – operating in only a few Carnegie Mellon buildings where wireless researchers were working.

I wondered if even the smaller network would work. I didn't know what problems lay ahead, but I suspected that the bad boys of radio were waiting to create some mischief.

((•))

Later the same day I walked across the patio at the end of the wide, grassy Mall and climbed stairs to the seventh floor of Wean Hall, a glaring exception to the university's yellow brick theme. Wean Hall recalled Soviet era high-rise architecture – concrete, austere, and imposing. I ascended the cinder block-enclosed stairwell, and, as I approached the seventh floor office of David Johnson, I wondered what Wean Hall's designers had been thinking. Like the stairwell, the building's corridors and offices had walls of undisguised cinder block.

A young computer science professor, David sat in his small office, surrounded by a random array of books, papers and reports stacked high. Every horizontal surface was covered. David was relaxed and dressed informally – he liked polo shirts and jeans. From the stacks of papers surrounding him, he could locate any document that he needed as part of our conversation. "Oh sure, I have it right here," David said. I was astonished. What looked like chaos was to him a well-ordered system.

David had created a project he called "Monarch" to build software that would be needed for mobile networking. The Monarch Project recognized the migratory – mobile – behavior of its namesake Monarch butterfly, but the name was also an acronym for MObile Networking ARCHitectures. David knew that a mobile computer, moving in space and connected by wireless, would, at any moment, have radio contact with some, but not all, of the other computers in a network. The computer-to-computer wireless communication links would naturally be established and broken frequently, depending on the exact locations of the moving computers. David was building the network software that would be needed in this environment – different than that used in a wired network, where communication links were well-established and reliable.

"I want a mobile computer – laptop, luggable or belt-worn – to seamlessly switch between data networks without any intervention from the computer user," I told David and asked, "Can you make this happen?" I hoped that he could write software that would allow a computer to use – as needed – any of a number of available wireless data networks.

David was enthusiastic. "Oh, sure. We can do that. No problem." His confidence made me think that it would actually happen.

I wanted a computer to be connected to the Internet through any wireless network – cellular and wireless LAN networks were two possibilities – with no need for the computer's user to worry about establishing, disconnecting or reestablishing the connection. It would be, as far as I knew, the first such software anywhere. And the system that David later created was much like the software that today allows an iPhone to switch between a Wi-Fi network and a 3G or 4G cellular network. It happens without any intervention from the iPhone's user.

Leaving David, I headed toward the Wean Hall office of the impeccably groomed and well-dressed professor, Mahadev

Satyanarayanan. I didn't know anyone who could actually pronounce his name, so we called him Satya, short for Satyanarayanan. Using a bit of geek humor, Satya once told me, with his faintly British-sounding accent, "the first five letters of my name – S-A-T-Y-A – carry all the information. The remaining letters are only for error checking."

Born to Indian parents living in Malaysia, Satya had attended the Indian Institute of Technology in the city of Madras and later received a Ph.D. from Carnegie Mellon. He stayed at the university to work on the Andrew project and then continued as a professor. Like David, he was interested in the new wireless network I was planning.

Satya worked on distributed file systems – systems that allowed data files to be distributed among the computers in a network. Your e-mail system is a simple example. Some e-mail messages, which are files, exist on your computer, some exist on the e-mail server, and some exist on both machines.

And Satya's work on distributed file systems made possible an early form of what we now call "cloud computing." User information – documents, spreadsheets, presentations, etc. – are stored somewhere in the network cloud, but the computer user doesn't know just where. When the user needs her information, it appears – as if by magic – on whichever desktop computer, laptop or smartphone is handy.

But, when files exist on more than one computer, it's important that all copies remain up-to-date and consistent. How can such consistency be maintained in a network with unreliable communication links – links that might be high-speed at one moment, much slower the next, and completely disconnected soon after that? Satya and his students were working to answer the question.

David and Satya both wanted to use the new high-speed wireless network to test their software. Other researchers were interested, too. Seven researchers in all wanted to use the new network.

I thought about which wireless technologies to use to build the new system. I knew that Bell Atlantic Mobile was beginning to experiment with Cellular Digital Packet Data – CDPD – to provide Internet service to users of mobile computers. But CDPD was slow. We might use it to provide off-campus service, but it didn't have enough speed for the campus wireless network.

And I thought about wireless LAN technology – the technology originally created to allow computer connections where building wiring was inadequate. I had read that network adapters would soon become available as PC cards – about the size of credit cards – that could easily be slipped into slots in laptop computers. Perfect! This was exactly what we needed to allow mobile computers to move around but still remain connected to the Internet.

I decided to use wireless LAN technology to build the on-campus wireless network and Bell Atlantic Mobile's CDPD service to provide connections to mobile computers that had moved off campus or outside the on-campus coverage area.

I needed to write a budget for the project, and the access points would be the biggest part of the budget. How many access points would be needed? It depended on how much area could be covered by a single access point. And that depended on the bad boys of radio.

Even murkier was the question of whether wireless LAN technology could someday be used to cover the entire Carnegie

Mellon campus. Could the bad boys be tamed to make this possible?

Like the Shadow and the Green Hornet of KOTZ's old time radio shows, I faced villains – bad boys – at every turn. As I moved along the path toward a wireless network, they hid behind bushes and waited at street corners, ready to pounce. I would need – not weapons – but a bag of technology tricks to fend them off. It would be a contest of wits – more like a chess game than an action movie.

Coverage would be a problem. The FCC didn't require wireless LAN systems to have a license because the equipment used low transmitter powers. But that meant there would be weak signals. I worried about achieving campus-wide wireless coverage, and I began to think about what I had learned in the past – from ham radio and from designing and building radio communication systems in Alaska – about the behavior of radio waves. Understanding radio waves' behavior would be the key to wireless LAN coverage.

A radio signal traveling from an access point to a computer's receiver would naturally become weaker as it traveled – even with no obstructing walls, floors, ceilings or office furniture nearby. The same is true in outer space. It's called *free space attenuation*. A weak signal becomes even weaker as it travels toward a receiver.

One of the bad boys lurking in the shadows would be *interference*. Weak signals would have to overcome the interference caused by other radio signals in the same radio band. Wireless LANs operated in unlicensed UHF bands, but unlicensed bands were a mixed blessing. The good news was that we didn't need an FCC license for our wireless LAN, but the bad news was that our neighbors didn't need one either, and their signals would interfere with ours. Other kinds of equipment in the same bands – cordless phones and others – would bring even more interference.

Another prankster was radio *noise* – natural or manmade. Radio noise occurs in nature – the static on your radio during a thunder-and-lightning storm, for example, but noise also comes from manmade sources. One is the buzz of fluorescent lights on your home radio. And, driving your car in a rural area, you may notice that a favorite station's signal competes with electrical noise radiating from your car's ignition system – or from the ignition systems of passing vehicles.

With our campus wireless LAN, there would be other noise sources to pose problems – malfunctioning electric motors scattered throughout university buildings were one. Microwave ovens were another. Would a wireless LAN signal arriving at a laptop computer be strong enough to cut through all this noise? It depended on how well we designed the system.

The other bad boys were the same ones I had faced while working on Alaska's bush telephone program – shadowing, reflection, refraction, scattering and diffraction. A wireless LAN's UHF frequencies were higher than the bush telephone's VHF frequencies. Some of the problems I had experienced in the Alaska bush could be even worse at the higher frequencies.

Shadowing would be a problem. A wireless LAN's signal could not penetrate a wall, ceiling or floor without being affected. The signal would be weakened. Two other imps were *reflection* and *scattering*. The wireless LAN's UHF signals would bounce off the building's flat surfaces, metal furniture and other objects. How much reflection and scattering would there be? It depended on the materials used in the building's construction, but I knew that metal objects like filing cabinets and desks in the buildings would cause lots of bouncing signals.

Some of the boys weren't all that bad. *Refraction* wouldn't be a problem over the short distances traveled by our wireless LAN signals – and *diffraction* might actually help a little by bending waves around the sharp edges of intersecting corridors to cover areas that would otherwise be shadowed.

But shadowing, reflection and scattering? Those guys would be crafty – and dangerous. Walls, ceilings and floors would shadow laptop computers from the signals being sent out by nearby access points. Reflection and scattering would bounce signals in all directions.

I thought about providing coverage to indoor areas by installing access points and their antennas outdoors, but I quickly realized that the shadowing caused by buildings' outer walls would make the already weak signals unreliable or unusable inside the buildings. We could use outdoor access points to provide coverage on the grassy areas of the Mall and the Cut, but, for indoor coverage, we would need to use access points installed inside the buildings. Years later, city governments of some major American cities made the same discovery when they used outdoor-mounted Wi-Fi access points to provide indoor wireless coverage to their citizens. The shadowing problem became obvious to the cities only after some of the systems were built and operating.

Radio signals from an access point mounted in a Carnegie Mellon hallway would bounce around corridors and rooms before being received at a mobile computer's wireless LAN receiver. After bouncing off walls, ceilings and floors, the signals would be recombined, meeting again at the mobile computer's receiver, and the result could be a stronger – or weaker – signal. This is the problem of *multipath*.

You may hear multipath as a flutter when you're driving you car and listening to an FM station. Or when you're waiting at a stoplight. First the signal is strong. You inch forward a bit. Now the signal is weak. The effect, caused by multipath, is the result of FM signals bouncing from nearby objects and recombining at the receiver.

Reflecting surfaces cause multiple copies of the same signal to arrive at a receiver, each at a different time. One copy follows a direct, straight-line path from transmitter to receiver, and others travel indirect, reflected paths, taking longer to reach the receiver. The delayed copies of the signal are like a multitude of echoes, and they arrive with different phase relationships to the direct signal.

Some of the reflections are in-phase, as when two pond ripples' peaks, or crests, meet each other, causing an even higher peak. These reflections make the direct signal stronger. But some of the reflections are out-of-phase, as when a pond ripple crest meets a trough. These reflections cancel out the direct signal, making it weaker. And when the receiver moves even a short distance the whole situation changes. New phase relationships among the direct and reflected signals cause rapid signal strength variations – from strong to weak, and from weak to strong.

((•))

A few months later, in March 1995, four of us sat around a long, polished oak conference table in Cyert Hall. The room's windows revealed a late winter day in Pittsburgh – cloudy skies and a light drizzle. Horns honked and traffic raced along Forbes Avenue, the busy thoroughfare on the north edge of the Carnegie Mellon campus. But we focused, not on the street outside, but on the whiteboard at the end of the room – it held our numerical ratings of wireless LAN products.

The project that had been only an idea 16 months earlier would soon be a reality. And the project had a name. I decided to call it "Wireless Andrew" because it would be a wireless extension of Carnegie Mellon's Andrew network, the one that was revolutionary in the early 1980s and later became the heart of the university's information infrastructure.

We were working on choosing a wireless product to be used in the new network, and our discussion was spirited. Several companies wanted to provide equipment. The competition was intense. We had narrowed the contenders to three, and this was the day for a final decision. We needed to move forward.

Each of us had a different way of speaking English, and we each had our own approach to evaluating the wireless LAN products. But we were close to a decision.

I sat at the head of the table. To my right was Ben Bennington, the new director of the Information Networking Institute. In late 1992 Ben had come from a federal government job in Washington, DC to take over the institute director's job that I had previously held. Full of enthusiasm, he was eager for the institute to begin the research initiative it had been lacking. "Let's get on with the meeting. We've done the analysis. Now it's time to make a decision," Ben said. His British accent lingered though he had left his native England many years earlier.

To my left sat Chuck Bartel, and at the far end of the table was John Leong. Both were directors in the computing services division that would someday be responsible for the operation of Wireless Andrew. As Carnegie Mellon's information technology organization, it provided computing and networking services to the entire campus.

A little younger than me, Chuck was a lifelong Pittsburgher and spoke with the accent we called "Pittsburghese." He had told me about spending almost all of his working life at "Kernegie Mellon," where he had risen from technician to director. Like all Pittsburghers, Chuck was a fan of the city's professional sports teams, including the Pittsburgh "Stillers." The rest of us – not native Pittsburghers – kidded Chuck about his accent. A good-natured guy, he laughed along with us.

John had joined Carnegie Mellon as a member of the original Andrew project team and stayed on, becoming a computing services director. John was a deep thinker. He would sometimes say, "Here's something else to consider" in the unusual accent that resulted from his growing up in a Chinese family and living in: Macao, a Pacific Island now a part of China; Mozambique, in southern Africa; and later in England, Canada, and the United States. John was a member of the technical staff, but he might have been a professor at the university. Professors are creative people – independent in their ways and sometimes a little eccentric. John was like that. He was an idea guy.

Several months earlier, I had received notification from the National Science Foundation that our proposal for funding had been approved. The agency had decided to give us $550,000 to design, build, and operate the new high-speed wireless network that would be a research test bed. Half a million dollars was a big chunk, and I wanted to spend it carefully because I doubted we would receive more funding from the agency.

After thinking about the practical details of carrying out the project, I had decided to enlist the aid of Carnegie Mellon's computing services division. Chuck and John, both directors in that division, became members of our team.

The Andrew network was once a research project, but, when it graduated to the university's mainstream system, it became the responsibility of the computing services division. Andrew became a *production network* – one that was expected to be reliable and dependable. I hoped that our new wireless network would someday be a production network, too – available to everyone at the university and run by the same computing services organization.

After we received the National Science Foundation funds, the four of us – Ben, Chuck, John and I – started work on choosing a wireless LAN product from the several that were then on the market. We contacted equipment makers and invited them to meet with us. Later we asked them to lend us samples of their products for testing. Some responded, and others didn't. In the end, we had serious discussions with three companies.

We developed a set of five criteria to use in evaluating the wireless LAN products. The five were: coverage, throughput, form factor, ease-of-use, and Apple Macintosh support. The last was important because, as Herb Simon had observed a year earlier, there were many Apple Macintosh computers being used on our campus.

Coverage was the most obvious of the five criteria. I thought about it frequently. We still didn't know exactly how much area could be covered by a single access point. The equipment manufacturers had given us some estimates, but we wouldn't feel confident about any estimate until we had done our own testing. We would draw conclusions from our own measurements.

Next was the problem of *throughput*, a measure of how much information could be sent through a network. The wireless LANs of 1994 and 1995 had advertised speeds of 1 to 2 Megabits per second, but these were peak speeds and didn't tell us how much data we could realistically push through the equipment.

Form factor – a bit of jargon – was just a fancy way of talking about the size and shape of the network adapter. Was it a PC card that would neatly fit inside a laptop computer? Or was it a much larger gadget – a "brick" or a "dongle" that would have to hang off the side of a computer? *Ease-of-use*, another question, was subjective, but that didn't prevent four engineers from assigning it a numerical value.

Ben suggested another criterion. We should try to buy equipment from a company that could be our development partner – not just our equipment supplier. Wireless LAN technology was developing quickly, and we needed to collaborate with a partner willing to take our suggestions about improving its wireless LAN product. We also needed a partner to advise us on building our wireless network. That's why we considered the companies' *partnership potential* as part of our deliberations.

We did elaborate testing. To measure coverage, we walked around, carrying wireless-equipped laptop computers to see how well they were able to receive signals at varying distances from access points. Bystanders were puzzled about why anyone would try to walk and, at the same time, stare at a computer screen. It looked dangerous.

To test throughput, we held a "bake-off." With an access point and several laptop computers in a room – each computer with an operator – all operators pushed buttons at the same time and began to move large volumes of data either to or from the laptops. We measured the elapsed time for each of these file transfers, and we did it with one, two or three computers operating at the same time, all competing for use of the network. The experiments were a geek's dream. They gave us objective coverage and throughput comparisons of the products we were testing.

Wireless LAN signals were weak, but they did have the advantage of being *spread spectrum* signals. By using a big slice of the radio spectrum, they could cut through noise and interference to get a message through. As a young ham radio operator, I had struggled, using all of the electronic tricks at my disposal, to hear my faraway friends' messages. These special signals might have made the job easier.

Spread spectrum technology artificially spreads a signal over a much wider range of frequencies than it would normally occupy. The spreading can improve the chances that even a weak signal will be received. Spread spectrum came in two flavors.

The technology was originally created for use in wartime, when an enemy might try to jam a signal. Causing a transmitter's frequency to hop around – first to one spot on the dial and then to another – in a seemingly random fashion, might avoid the enemy's jamming signals. The trick was to move around, using a secret sequence of frequencies, faster than the enemy could follow along. If the transmitter and receiver hopped together, using the same sequence of frequencies, it was possible to avoid the interference caused by an enemy jammer. But the method used a big chunk of radio spectrum.

This original way to create a spread spectrum signal was called *frequency hopping*. The technique was first mentioned in a 1900 patent filed by radio pioneer Nikola Tesla. The U.S. Army Signal Corps used it during World War II to provide secret communications between President Franklin Delano Roosevelt and British Prime Minister Winston Churchill. That system was top secret and remained so until the 1980s.

And the glamorous and seductive actress Hedy Lamarr was – surprisingly – a radio inventor listed on a 1942 patent. Her story was strange one.

Born in Vienna, Austria in 1914 as Hedwig Eva Marie Kiesler, she quit school at 16 and became a film star – once appearing nude in a film – and then married an older man. During her marriage to arms dealer Friedrich Mandl, he took her to meetings with engineers and business partners, where she learned about military technology. Otherwise, Mandl tried to keep her shut up in his mansion. When she discovered he was cooperating with the Nazis, she drugged him and, with the help of her maid, escaped on a train to Paris. Her personal story had a better plot than some of her movies.

Eventually, Hedy made her way to Hollywood, where, using the stage name Hedy Lamarr, she became world famous. She made 18 films from 1940 to 1949, and, in the 1940s, her name was a household word. She appeared in films with Bob Hope, Victor Mature, Lana Turner and Judy Garland, among others. But not so well-known was her work as an inventor.

In Hollywood she met George Antheil, a musician and inventor who had experimented with automatic control of musical instruments. He had written *Ballet Mecanique*, a ballet written to be played by those old-time player pianos controlled by rolls of paper peppered with holes. Each hole represented a certain note to be played at a certain time. The *Ballet Mecanique* was performed by a number of the pianos playing simultaneously and in synchronization. It was a complex combination of musical instruments.

Remembering the military technology meetings she had attended, Hedy knew that, if a torpedo were guided by radio signals, it could more effectively hit its target. But she also knew that the radio control signals could be jammed by the enemy. Hedy worked with George to create a frequency hopping method that used the holes in a piano roll to control a radio's frequency hopping sequence. Each hole represented a certain radio frequency to be transmitted at a certain time. Using 88 different frequencies – the same as the number of keys on a piano keyboard – the invention would make the signals harder for enemies to jam.

Although the patent was issued in 1942, the mechanical technology of the day was not up to the job, and the pair never made money from the idea. But Hedy Lamarr was later honored as an inventor and a pioneer of spread spectrum technology.

And there is another way to create a spread spectrum signal. Called *direct sequence* spread spectrum, it converts a message's string of ones and zeros to a much longer string. The method works because the spectrum occupied by a radio signal is directly related to transmission speed. Transmitting the longer string of ones and zeros in the same time as the original string means faster transmission and causes the signal to occupy more spectrum. The method is more complex than frequency hopping, but its advantages are similar.

In 1995 some wireless LAN products used frequency hopping spread spectrum, and others used direct sequence. Of the three products we seriously considered, two used frequency hopping, and the third used direct sequence. Each company claimed that the method it used was better than the other.

I had done my own analysis and decided that, with the power limitations imposed by the FCC, direct sequence spread spectrum would give greater range than frequency hopping. My conclusion was borne out by our testing. In fact, the ranges we measured with the direct sequence product were similar to the estimates I had made in writing the project budget. That was good news. The test results suggested that we could use the same number of access points the proposal had assumed. And the test results also told us we should consider using a direct sequence product.

Years later, the wireless LAN industry seemed to agree with our spread spectrum conclusion. When Wi-Fi standards were adopted, they at first included both direct sequence and frequency hopping spread spectrum. But direct sequence proved to be far more popular, and soon it was used by most Wi-Fi products. Later versions of the standard completely eliminated frequency hopping.

We chose the wireless LAN product sold by AT&T. The product, called WaveLAN, had been originally developed by NCR – National Cash Register – before that company was bought by AT&T. The WaveLAN product scored well in all of our categories. Its direct sequence technology gave it greater range, and we thought that AT&T would be a good development partner. We liked their development group, located in Holland. The company was large, and it seemed stable.

In March 1995, with our National Science Foundation funding approved and our wireless LAN product chosen, we were ready to begin a trans-Atlantic partnership with our new Dutch colleagues.

CHAPTER 5

THE DUTCH CONNECTION

MAY 1995

A stream of students rushing to the day's first classes swept me across the campus. Spring's morning dew glistened on the wide swath of lawn we called The Cut. Trees showed tiny buds.

I looked for student news on The Fence, a sort of bulletin board for fraternities and student organizations. Its once slender posts and rails were fat with paint – layers of long forgotten announcements advertising everything from weekend parties to charity fundraisers. According to the rules, the Fence could be painted each night, but only between midnight and sunrise. Students warily guarded their artwork through the night to prevent others from painting over it. At just before eight in the morning, long after sunrise, the overnight bulletins were safe, but a young guy was still on guard with his boom box loud. Madonna sang "Take a Bow."

I strode past the big antique clock, sitting on a high post ticking off the seconds, and entered Doherty Hall, a yellow brick building built in the university's traditional *beaux arts* style. The wide corridor's downward slope pushed me forward. I smelled the building's pungent chemistry labs. At the end of the cavernous hallway, I loped down a flight of stairs to the first basement level and strode toward an indoor connector link leading to the neighboring Wean Hall.

Stepping off the elevator on Wean's top floor, I glanced at my watch. After rushing to be on time for a meeting, I was surprised to be early. But it was a chance to look out at the panorama encircling the building's highest floor. A window at the end of one corridor gave a clear view to the northwest – the nearby University of Pittsburgh campus, dominated by its 42-story Cathedral of Learning. The towering structure dwarfed its surroundings. Said to be the highest building on any U.S. university campus, the gothic monolith can be seen for miles in every direction. It's a beacon that guides Pittsburghers navigating the maze of streets that sprawl over the city's hills.

Through the open door of a professor's office, I could see the pastoral scene to the south. A patio separated Wean from Baker Hall, one of Carnegie Mellon's original yellow brick buildings, and past Baker was the manicured Schenley Park. In the distance lay the Phipps Conservatory, a grand glass structure housing an array of plants from around the world. The magnificent building shimmered like a crystal palace.

Waiting for the others to arrive, I daydreamed about the journey I had made from my teenage ham shack to a small Eskimo village and then to a world-class university. I thought about my radio experiences, all with analog radio. The new wireless network, using digital radio, would be a first for me.

My old ham transmitter had been analog. It used AM – amplitude modulation. From one instant to the next, the transmitter's output, its amplitude, varied continuously, in step with the air pressure variations of the sound waves reaching its microphone. Hundreds of times each second, the pressure went through peaks and valleys, and the variations were tracked by the transmitter's output. In other words, this output was an *analog* of the air pressure at the microphone. The Alaska radio stations – KOTZ, KSKA and others – had all used forms of analog technology.

But the digital radio signals used to send binary information – information comprised of ones and zeroes – do not vary continuously. Unlike their analog brethren, they may assume only certain, predefined levels. Any information – speech, video or data – can be reduced to binary ones and zeros and then communicated using only the allowable levels of a digital signal.

I thought about how I could have converted my analog ham radio transmitter to digital, permitting it to send ones and zeros. Each time a binary one appeared at the transmitter's input, the transmitter would be turned on, and, for a binary zero, the transmitter would be turned off. A one would be converted to power on, and a zero would be converted to power off. It would have been the simplest kind of digital radio.

A special receiver would be needed to pick up and decode the digital signal, converting it to the original string of ones and zeros. The receiver would examine the incoming signal in each time interval and decide if it looked more like a one – transmitter on – or a zero – transmitter off. No incoming signal, or just a weak incoming signal, would look like a zero. But a stronger signal would look more like a one. The receiver would be guessing if a one or a zero had been transmitted. It would be an educated guess – but still a guess.

But there could be problems. Noise or interference could make it appear, even when no signal had been sent by the transmitter – a zero – that there was a signal present – a one. The receiver could make a mistake. And noise or interference could make a transmitted one look like a zero, another kind of mistake.

Though most digital radio systems are more complex than my simple, imaginary set up, they use the same ideas. A transmitter converts ones and zeros to create certain, predefined signals, and a receiver guesses whether ones or zeros were transmitted. Weak signals, noise and interference can cause a digital receiver to become confused and make errors – incorrect guesses.

Wireless Andrew would use digital radio for communication between computers. Sound, video and other information would be sent digitally after being converted to ones and zeros. But the wireless network, using only low-power because it was not licensed by the FCC, might have problems. If the received signals, after being weakened by shadowing, weren't strong enough to overcome interference and noise, errors would result. If there were too many errors, Wireless Andrew wouldn't work, and the bad boys would be the winners.

I didn't really know if our low-power wireless network would have the strength to overcome interference and noise. I had taken a risk. In late 1993, when Marvin and I looked for a way to jumpstart our research program, the wireless network seemed like a good idea, but now I wasn't sure if it would work.

Writing the National Science Foundation proposal and budget wasn't hard. I had estimated that one access point, a wireless base station, would cover a certain area in a building, and this made it easy to calculate the number of access points we would need. We had also done some coverage testing to try to verify these estimates. But were our figures correct? Would we have enough money to buy all of the needed access points? I didn't know. No one did. A network like this had never been built.

We decided to test the estimates by designing a network to cover the top three floors of Wean Hall, where some of our wireless researchers worked. We would need to choose access point locations that would give complete radio coverage to all three floors. It would be a challenge.

Carnegie Mellon network technician Mark Campasano startled me when he rounded a corner, leading two AT&T technicians toward me. "Camp, how's it going?" I asked. Mark replied, "No problem so far." He was a man of few words.

It was the first day of our network design work in Wean Hall. I wanted to be sure that Mark and the two AT&T technicians had all the equipment and information they would need to do the design. I wanted things to go well. Making quick introductions, Mark told me the two technicians' names. I told them mine and shook their hands. Mark said they had all the equipment they needed and a set of Wean Hall drawings. The three men were ready for their task.

Mark was becoming a quiet hero of the Wireless Andrew project. Born to an Italian family in Pittsburgh, he had always lived in the city. Solidly built and with dark hair, he was a quiet guy, but behind those bushy eyebrows was a quick mind. He was the perfect network technician to work on this first design effort, and I knew there was no need for me to hang around. "Good luck," I called out, heading toward the elevator.

Crossing the patio, I walked toward my Baker Hall office, still thinking. Designing Wireless Andrew wouldn't be easy. The big problem was coverage. We needed to arrange the access points so that, even with the problems their radio signals would face, signal quality would be good enough to ensure that those ones and zeros would be received correctly. If the access points were placed too far apart, there would be dead zones, where signals wouldn't be strong enough for correct transmission. If access points were placed too close together, we would use more access points than needed, wasting precious money.

It was a "Goldilocks" situation. We wanted the access points not too close together and not too far apart. We wanted them just right.

The Wean Hall design would be a demanding task because of the building's cinder block, steel and concrete construction. Signals would not easily penetrate the building's walls, floors and ceilings. Instead, they would bounce from these surfaces in a complex jumble of lines and angles. Yet, if we could successfully build a network in Wean Hall, we could probably do the same in any other Carnegie Mellon building.

We had asked AT&T, our new partner and equipment provider, to help us design the Wean Hall network. Working with Mark, the company's two technicians soon finished making measurements. Preparing to leave for the airport, they promised to send us a plan telling us exactly where to position the access points. But when we received the AT&T report a few weeks later, Mark did more testing and found that the design had coverage gaps – dead zones. I didn't like the news. We needed to do another design before installing access points.

The managers at AT&T agreed to send the team back to try again. The same two technicians returned in July 1995, only a few months after their first visit, to repeat the measurements. They revised the Wean Hall design, and they also created wireless designs for future wireless networks in a few other buildings. The new designs looked better. We were ready to install access points in Wean Hall.

((•))

Our first contact with AT&T had been a year earlier, when we were trying to decide which wireless LAN product to use in our network. We knew that several companies manufactured and sold wireless LAN equipment, but all of the products were meant to be used in a small coverage area, usually to interconnect computers within a single room. I wanted something bigger – a wireless network covering the entire campus. Still, the wireless LAN products were a good starting point.

One of the wireless LAN products on the market had been developed by a small group in a small country. They were in Utrecht, a city in the Netherlands, and they wanted to sell us their equipment. The little Dutch team had energy and passion that rested on a belief in what they had developed.

We first met Cees Links, the group's product management director, when he traveled to Pittsburgh to tell us about the product that they called WaveLAN. Cees – pronounced "Case" – was young, dapper and articulate, and he spoke perfect English – complete with slang. Like many of his countrymen, he was also fluent in French, German, and his native Dutch, the result of growing up in a small country sandwiched between bigger ones speaking other languages.

Cees was a good salesman, and, more important, we liked the WaveLAN product. Our technical analysis showed that it was the best one for our project. Still, its low-power, limited-coverage equipment was not meant for an ambitious campus-wide network like the one I wanted to build.

We didn't know it at the time, but Cees and the team in Holland were, like us, thinking about how wireless LAN coverage might someday be expanded beyond one or two rooms. In describing the team's thoughts in this time period, Cees later said:

> In principle the WaveLAN card was intended to replace the cable between the PC and the main computer or central server in the computer room. However, in many cases the computer room was too far away, even on another floor. We developed many ideas to solve this distance problem: repeaters, leaky cable, passive antenna amplifiers. None of them was considered feasible except for the concept of the station, initially called access point (AP). The functionality of the AP includes transmitter, receiver, antenna and the bridge to route packets to and from the wired network infrastructure, essentially the Ethernet. We realized that we would need multiple APs to cover larger buildings, to be connected to the wired LAN infrastructure; plus the capability of roaming between the APs. When walking around in the building with a portable/laptop you should remain connected and when coming within the range of another access point the communication session should not be interrupted as result of the switchover.[1]

---

[1] W. Lemstra, V. Hayes and J. Groenewegen, *The Innovation Journey of Wi-Fi*, University of Cambridge Press, New York, 2010

The Utrecht engineers thought it might be possible to use many access points to create a large wireless LAN coverage area, but they had no way to do *handoff* – they called it "roaming" – transferring a mobile device's connection between access points as the user moved from place to place. Cellular phones could do this but wireless LAN products couldn't.

And there was another problem. The Dutch group didn't have a good plan for how to place access points inside buildings or in outdoor areas. A design method was needed that would allow wireless service over a large area like a university campus. We thought that, as AT&T worked on the handoff problem, we could work on how to place the access points. It could be a good partnership.

((•))

With AT&T's second Wean Hall design in hand, we were ready to install access points. In 1995 each access point needed two kinds of wiring, one to make connections to the wired campus data network and the other to provide electrical power. Our own network technicians could do the communication wiring, but the power wiring would need to wait for the licensed electricians who worked for Carnegie Mellon's facilities department. We placed orders for the power wiring and then waited…and waited…and waited.

A few months later, we finally received word that all of the power wiring was complete. Our network technicians quickly did the communication wiring and installed the new access points. We were ready to try out the Wean Hall network.

But there was more bad news. When we turned on the equipment and took measurements, there were still dead zones. Many of those ones and zeros were not being correctly received. The network wasn't working as expected. It was a setback that seemed to raise the old questions anew. Was it possible that the low-power signals, after shadowing and other problems, were just not strong enough to overcome the interference and the noise in Wean Hall? Could we really expect to use this technology for a multi-building network?

I told Mark that we shouldn't have been surprised. After all, no one had ever built a network like this. No one yet knew how to design a large-scale wireless LAN. As I reassured Mark, I was also trying to reassure myself.

Our problem was the bad boys were hanging out in Wean Hall. Wireless signals were being shadowed by walls, doors, ceilings and floors, weakening the signals and rendering them helpless to enter other levels of the building – or even other rooms. Metallic surfaces – metal doors and metal file cabinets – were like mirrors, reflecting the signals. Textured surfaces like cinder blocks scattered the signals in all directions. Signals bounced everywhere in a crazy quilt of lines of angles.

And copies of the same radio signal bouncing from walls, floors, and ceilings could meet again, forming a new signal at a receiver's antenna. But the resulting signal might be stronger or weaker than the original. Just as pond ripples crossing each other can, at any moment, make a bigger or smaller ripple, the radio waves recombined to make a received signal strong one moment and weak the next. The fast variations in signal strength, like the ones sometimes heard on an FM car radio, were caused by multipath.

There had been similar challenges in building Alaska's bush telephone network, but only a few objects in each village threatened to cause shadowing, reflection and scattering. An Alaska village might have a hilltop or a few rooftops that could cause problems. It was a manageable situation. But how we could deal with the complexity of Wireless Andrew's shadowed, reflected and scattered signals, a jumble of lines and angles stretching up and down Wean Hall's corridors?

Our colleagues in the wireless group at Virginia Tech were working on using a computer to analyze this kind of situation. They wrote a program that could calculate all of the angles and bounces in order to estimate signal coverage. But, to work properly, the software needed an exact physical description of a building – floor plans, ceiling and wall thicknesses, building materials, and other information. I decided that we shouldn't try to analyze such a complicated web – with or without a computer. It would be better to just accept the end result. In fact, there really wasn't a choice because our building drawings didn't have the detailed building information needed by the Virginia Tech program.

I became more confident that this was the right decision when Mark made a surprising discovery. He found an access point on Wean Hall's eighth floor that had no coverage at all on the seventh floor, but its signal appeared – magically, it seemed – on the sixth floor, providing coverage there. We wouldn't have predicted this, but it was easy to measure.

A few years later, reflecting on our wireless work, I wrote:

> ...experience shows that the [access point] layout must be based on measurements – not just on "rule of thumb" calculations. These measurements involve extensive testing and careful consideration of radio propagation issues when the service area is large, such as an entire campus.[2]

---

[2] A. Hills, "Large-Scale Wireless LAN Design," *IEEE Communications*, vol. 39, no. 11, pp. 98-104, November 2001

In other words, don't try to predict coverage – just measure what you have, and design accordingly.

We worked out a way to choose access point locations by making signal strength measurements. In each building we made hundreds of measurements, placing access points where we thought they might work, measuring their coverage, and then trying new locations. It was time consuming, but it worked.

But coverage wasn't our only problem. With the Wean Hall network up and running, we began to notice, even when coverage was good, the problems of the congestion caused by too much data flowing between computers. If there were too many computers using one access point, it could become overloaded, causing delays that computer users perceived as slow service.

The problem was that, with all access points and computers using the same radio channel, only one station, an access point or a mobile computer, could successfully send at a time. It was like truckers using a single CB channel. If two of them transmitted at the same time, both messages could be lost in the jumble of interference.

AT&T's WaveLAN equipment resolved such conflicts by using a technique called "carrier sense multiple access with collision avoidance," a fancy term that could be explained simply. Before sending, a polite CB operator waited until others finished their transmissions. With WaveLAN and other wireless LAN equipment, each station listened before sending, waiting if another was using the channel and transmitting only when no other station was on the air. All stations followed the same strict radio etiquette.

But, if two stations attempted to send at the same time, neither would hear the other, and their transmissions would collide. The two transmissions would interfere with each other, neither would be received correctly, and repeat transmissions would be needed. With many computers using a single access point, these collisions occurred frequently, too many repeat transmissions were needed, and the network slowed down.

We noticed times when the network was slow, and we understood what was causing the problem. But we didn't immediately work on a solution. Coverage was a more important concern.

By late 1995 we had wireless working in six buildings. In spite of a few remaining problems, researchers began using the network. They loved it. Things were looking up.

Still, we saw things we wanted to change. We needed to make our design better. We were in uncharted waters and couldn't expect to get everything right the first – or even the second – time. To improve on what we had done, we decided in the summer of 1996 to completely redesign the six building network, using some new ideas, and to add a seventh building, the ivy-covered Baker Hall.

And reinforcements arrived in September 1996, when Albert Eikelenboom stepped off the plane in Pittsburgh. A technician and a member of the AT&T team in Holland, Albert had come to help us make more measurements and improve our design process. Albert was one-of-a-kind – a memorable character and a guy we all looked up to – literally. I was six feet four inches tall, usually towering over others. But Albert was a different story. Nearly seven feet high, he towered over me. He didn't outweigh me, though. Albert was a beanpole.

And he was earnest. He listened intently to our descriptions of the project and our concerns about the network. With a seriousness of purpose, Albert proposed and then executed a work plan to address our problems. He finished by writing a detailed report, explaining his findings and recommendations.

Albert was contemplative both on and off the job. He was a deep thinker. We weren't surprised to hear that he was also serious about his religion or that he was intensely involved in his home church in Holland. But we were startled to hear that, a few months after his return home, Albert planned to participate in his church's Christmas celebration by playing the role of "Father Christmas" – something like our Santa Claus. But the seven-foot beanpole didn't match our image of a jolly old elf.

After his return to Holland, we received Albert's detailed report on what he called the "measurement campaign." His careful and meticulous work was a key to refining our design method, and, as we read the final report, we thought of the Dutch Christmas celebration with the thinnest Santa anywhere.

During the fall of 1996, we worked in the first six Wireless Andrew buildings, moving access points and installing some new ones according to our latest design. We also installed new access points in Baker Hall, the newest Wireless Andrew building. In January 1997, a check of all coverage areas showed that the network was operating just as we had hoped.

We soon decided to "release" the seven-building network. This meant that its operation would become the responsibility of the university's computing services division, and it would be available to be used by any wireless researcher. Three years into the project, we had reached a milestone, finally doing what we had promised the National Science Foundation. We had built a high-speed wireless network, and researchers were happily using it.

But I was beginning to hear that there were others on the network.

CHAPTER 6

STUDENTS LIVING IN THE FUTURE

OCTOBER 1998

The stern, larger-than-life professor stared straight ahead, draped in academic colors. He seemed not to notice the young guy in jeans and a T-shirt sitting on the floor before him. The student didn't notice the professor either. He was intent on his laptop computer, holding it above his head, pointing it first one way and then another.

I stopped to talk with the student. We were both overshadowed by the professor's big portrait, which dominated the room.

The recently arrived freshman told me he was holding his laptop high so that its clip-on video camera could take in a sweeping view of the building's lobby. The laptop was linked via Wireless Andrew and, through the Internet, to a distant friend. At the other end of the link, his high school classmate, now an MIT freshman, watched the impromptu tour. The Carnegie Mellon student was proud of the wireless network at his new university. He wanted to be sure that his MIT friend knew about the "cool" new service, one not available at MIT. And he wanted to rub it in.

The Wireless Andrew network was supposed to be available only to Carnegie Mellon wireless researchers – not to students. We had created this policy to prevent the network from becoming overloaded, disrupting the researchers' work. Software controls enforced the policy, but they were no match for a smart Carnegie Mellon student – even a freshman.

Other chance encounters with students had already told me that lots of them were using Wireless Andrew. I began to think that we were working on something more than just an interesting research project. Our students were often the first users of a new technology, and they were a reliable leading indicator of interest in future products and services. It looked like there might soon be a big market for wireless networks just like the one we were building.

I looked up at the man inside the picture frame. He wasn't just a professor. Former president Richard Cyert had led Carnegie Mellon from 1972 to 1990, transforming it from a regional engineering school to a world-class university. He had presided over the inception of new interdisciplinary program areas – urban and public policy, social history, engineering and public policy, and cognitive psychology, among others. And he created an environment that nurtured innovative projects like Wireless Andrew.

And Richard Cyert's successor, Robert Mehrabian, became a Wireless Andrew advocate. When he learned of our little wireless project, Robert became an instant enthusiast. During his presidency, he showcased the Wireless Andrew network, intrigued that it could give faculty, students, and staff freedom to teach, learn and communicate anywhere on campus. Given the chance, he happily described Wireless Andrew to anyone who would listen.

But my thoughts drifted to Carnegie Mellon's first president, Andrew Carnegie himself. His big portrait hung on the top floor of the university's central administration building. The Carnegie in the painting looked regal, gazing out from behind a full gray beard. He was a wealthy and successful entrepreneur. He was Warren Buffett and Bill Gates in one, richer than both put together.

Carnegie had arrived in the United States as a child with his Scottish parents. He started working as a messenger boy and soon rose to telegraph operator. From these beginnings, he became a captain – no, an admiral – of industry. His industry was making steel, and his fortune was equivalent to 300 billion modern dollars.

Yet Carnegie was best known as a philanthropist. He was founder of the Carnegie Corporation of New York, the Carnegie Endowment for International Peace, and the Carnegie Museums of Pittsburgh. He founded the Carnegie Technical Schools – now called Carnegie Mellon University – in the year 1900. By the time he died, he had given away most of his money to establish libraries, schools and universities across America. At Carnegie Mellon, we lived by the founder's words, "my heart is in the work." He was a workaholic, and so were we.

I left the freshman, still sitting under the Cyert portrait, and walked toward my new office on the first floor of the former president's namesake building. I was now head of the university's computing services division, located in Cyert Hall.

As the university's chief information officer, I was responsible for the university's computing, telecommunications and network systems. It was a tall order. Carnegie Mellon faculty and students were demanding. They expected computing and network services to be the best at any university and to work perfectly all the time. I was grateful for the 150 employees working in the computing services division. They were the people who kept things running.

My appointment to the new job was good for Wireless Andrew. I wanted the network to evolve from a small one aimed at supporting a group of wireless researchers to a full-blown network serving the entire university community – professors, students and staff. And the campus-wide system would need to become a production network, providing the same high-quality reliable service as other systems operated by the computing services division I now headed. My new job would allow me to move the project in that direction.

I strolled through the waiting room outside my office. The receptionist was busy with a phone call, and I greeted her with a wave. Before my first meeting I had a few minutes to catch up on memos and e-mail messages. Yet, sitting at my desk, I was distracted by the cars, trucks and motorcycles speeding along Forbes Avenue, just outside my window. It was a gray October day, typical for Pittsburgh, and the traffic's buzz was constant. Students darted across the avenue, dodging buses, trucks, and cars and rushing to reach class on time.

Crouched beside Forbes Avenue, Cyert Hall had an odd, glassy look. It was another of Carnegie Mellon's misfits. Hugging a row of small maples lining the street, it was hemmed in by neighboring campus buildings. Its boxy shape and big glass façades didn't match the university's architectural theme.

My new office was the same room where our crucial March 1995 equipment selection meeting had taken place – the meeting where Ben Bennington, Chuck Bartel, John Leong and I had selected AT&T as our equipment provider and wireless partner. John later left Carnegie Mellon to join a Silicon Valley startup company, but Ben and Chuck remained to become key players in the Wireless Andrew project. With my new responsibilities, I needed some help. Ben and Chuck teamed up to take on day-to-day supervision of the project.

As director of the Information Networking Institute, Ben became, like President Mehrabian, a Wireless Andrew enthusiast. He talked up the project, telling researchers with his British accent that they should try the new wireless network. He told outsiders that we were creating a new model for mobile computing. Ben knew that Wireless Andrew was a creature of the institute he now directed, and it could put the institute on the research map.

Chuck was both amiable and unflappable. He was a director in the computing services division and a natural to lead the design and installation effort. Chuck had worked at Carnegie Mellon for a long time. He knew how to keep the design and installation effort moving forward.

Sitting at my desk, I noticed that the clouds had parted. Sunlight touched the cars zipping along Forbes Avenue. But it was time for my meeting. I left the daylight behind, walking back through the waiting room and down the hall to a conference room with fluorescent lights in the ceiling but no windows to admit sunlight.

Others drifted into the room in ones and twos, cradling coffee cups. We clustered around the big pot of coffee on the conference table and began an intense session, poring over the building floor plans spread out before us. We had done this many times, at least once for each building's wireless design. I looked around the table. We had a great team. Andrew Carnegie would have approved.

Our newest team member was Lisa Picone, the first woman to join us. She was born and raised near Pittsburgh, but, because of her travels, her Pittsburgh accent had faded. She had been an electronic technician in the U.S. Navy, making the rank of Petty Officer Second Class and spending most of her eight Navy years in Europe, some in Naples, Italy and the rest at SHAPE – Supreme Headquarters Allied Powers Europe – in Mons, Belgium.

Young and attractive, Lisa still worked easily with the guys. She was, after all, a Navy veteran. She didn't need to draw attention to herself. She was quiet, confident, and capable. She knew what she was talking about.

Lisa had joined Mark as a technician on our field measurement team, working in buildings across the campus, doing coverage testing and choosing access point locations. Her low-key style helped us in unexpected ways. More than once, she was doing measurements in a university building when a departmental secretary worried about the new wireless network and asked if the access points would cause cancer. There had been news coverage about possible health effects of cellular telephone systems. Cellular was a different kind of wireless, but it was still a reasonable question.

Lisa patiently explained that the energy exposure from an access point was far less than from a cell phone. And there was no research evidence that either Wi-Fi or cell phone signals caused any kind of health problems at all. Expressed in these rational terms, the argument was persuasive, but it was really Lisa's gentle, low-key style that made the difference, relieving employees' concerns.

In the windowless room, sitting around the conference table, we, one by one, fell silent, descending into thought. We were following a familiar routine, using a wireless design process that was by then well developed.

Two years earlier we had sketched the beginning of a design method and, over time, improved it bit by bit. We aimed to place access points in a way that would provide complete coverage. Our design process assured, even with fluctuating radio signals, that a computer user anywhere in a building would have a reliable network connection. The key was correctly locating the access points.

Yet our conference room design sessions were possible only because Lisa and Mark had spent many hours making measurements. The two worked to find the best places to install access points, and their task was complex. The coverage area of one access point interlocked with the coverage areas of other nearby access points. All of the access points worked together.

In each building the two technicians went through a step-by-step process. They chose a location for an access point, located a second access point, and then a third. They didn't try to analyze the jumble of radio signals bouncing up and down the hallways – they just measured signal strengths and placed access points accordingly.

Sometimes I worked with Lisa and Mark, helping with the measurements. They followed what we sometimes called the "duct tape method." It was a trial and error process. Mark set up an access point in a temporary location, using duct tape to hold it on the wall. Lisa measured its coverage by walking through the building and measuring signal strengths to find the edge of the space where a good signal could be received. Then Mark chose a location for a second access point, a location that would give complete coverage of a second part of the building, and Lisa made more measurements. It was a sequential process. First one access point…then another…and then another. It sounded simple, but, with all its details, it wasn't.

It often didn't make sense to place an access point either directly above or directly below one already installed on an adjacent floor because their coverage areas would have too much overlap, causing unwanted interference between the two units. So we staggered the access points, moving the second access point down the hall so that the coverage areas of the two access points, one on the floor above and the other on the floor below, dovetailed. It was time consuming, but the two technicians were patient. And patience was valuable – especially on this project.

In our conference room meetings, we worked to create better, more effective techniques for Lisa and Mark to use. We developed the procedure that has since been used to design many Wi-Fi networks across the US and the world, but then the process was a new one.

On the big conference table were a set of building drawings that Mark and Lisa, working from their measurements, had converted to coverage maps. They used colored shading to show the coverage areas for each building. Blue showed the coverage area of one access point, red the coverage area of another, and green another. The colors made it easier to visualize the access point coverage areas in the three-dimensional space of a building.

Sitting around the table in deep thought, we all tried to think of ways to improve the technicians' design by adjusting the access points' locations. Sometimes there was a better idea about their placements. After everyone agreed on a design, it was approved by the team, clearing the way for access points to be installed.

But that still left the problem of channel assignments. We needed to decide which radio channel would be used by each access point because our network now had three channels. Traffic congestion, and therefore performance, of the network depended on making these channel assignments correctly. This additional task was needed because of an event that had happened a year earlier.

((•))

1997 was the year that the standards organization of the Institute of Electrical and Electronic Engineers – the IEEE – adopted its new IEEE 802.11 standard. It was the first standard for wireless LANs and soon became known as "Wi-Fi," a name that was better for marketing than a string of letters and numbers.

Wireless insiders knew that the standard was a big deal. It assured a purchaser of wireless networking equipment that it would be compatible with similar wireless networking gear made by other manufacturers. This hadn't been true before 1997. Our original AT&T WaveLAN equipment, for example, was incompatible with equipment made by other wireless LAN manufacturers. A network user with equipment made by a different company could not have used Wireless Andrew in those early days.

But the new standard changed things. It gave customers confidence that the new wireless technology was not just a fad. Equipment manufacturers were committed to making compatible products. They believed in the new wireless LAN market. The standard told the world that the new technology was real and was here to stay.

However, the standard meant we would have to move Wireless Andrew to a higher radio frequency. We had been using the 915 MHz UHF band, a band that was unlicensed in the United States but not in many other countries. Wi-Fi equipment would use the higher 2.4 GHz UHF band, one that could be used for unlicensed operation in almost all nations. Companies making the new 2.4 GHz Wi-Fi equipment could confidently expect to sell their products around the world.

We were happy to abandon the 915 MHz band. Spectrum congestion and interference had been causing problems for us. There were too many other signals at 915 MHz – everything from cordless telephones to baby monitors – because no FCC licenses were needed. We called it the "kitchen sink" band – everything was in it save the kitchen sink. And, to make things worse, there was a high power paging tower just past the University of Pittsburgh's giant Cathedral of Learning. The tower's antenna, as it sent signals to personal pagers, or "beepers," interfered with our 915 MHz signals.

The 2.4 GHz band was more spacious than 915 MHz. It was a bigger chunk of radio spectrum. 2.4 GHz allowed the use of three Wi-Fi channels – not just the one we had been using. And this permitted *frequency reuse*. We could set access points to use different radio channels, minimizing interference between access points and providing much better network performance.

It was the same technique that had been used for years by the cellular telephone industry to efficiently utilize available spectrum. The technique could now be used with Wireless Andrew, but it would complicate our design process.

And 2.4 GHz came with its own set of problems. It had a different kind of unlicensed interferers – the ones used to cook food. Microwave ovens heat food by exciting the water molecules they contain, causing the molecules to do a little dance, increasing the intensity of their already frenetic motion. This increases the temperature of water – and the foods that contain water. And microwave ovens operated at 2.4 GHz, the same frequency used by the new wireless standard. But some microwave ovens were more troublesome than others.

Our experiments found their differences. The mesh encased in the glass of an oven's door is part of the shielding that prevents radio energy from escaping into the surrounding area. New ovens were pretty well shielded. It was the older, leaky ones that allowed radio emissions to escape, and these were the ones that caused interference. Over time we weeded out the troublemakers on campus and then didn't have further problems with interfering cookers.

And 2.4 GHz cordless phones had begun to appear – another source of radio interference for Wireless Andrew. In the end, these phones would be a bigger problem than microwave ovens.

But, overall, the new standard was a positive development. It gave me confidence that Wi-Fi equipment would be a market success. It seemed likely that other universities, companies and government agencies would soon follow our lead and begin to build Wi-Fi networks like ours.

Watching our students use the new technology told me that the demand for Wi-Fi would be big. Wireless Andrew had attracted lots of unauthorized users like the freshman sitting under President Cyert's portrait. I received messages from many of them, asking how they could sign up to use the network. The official answer to this question was always the same: "Wireless Andrew is a research network and is intended only for the use of faculty members doing research in the wireless area."

But Carnegie Mellon students are clever, and they found ways to connect to the network. I knew what was going on and didn't say much about it. Technically lawbreakers, the students were really "early adopters" and up-and-coming technology innovators. I didn't want to discourage them.

It seemed that the vision of anytime-anywhere high-speed wireless Internet access could soon become reality. I thought about expanding Wireless Andrew to cover the entire campus – my original dream – and opening the network for use by the entire campus community. The new Wi-Fi standard was one piece of what I needed to make the big leap. But I also needed money – or its equivalent. It was time to talk to our partners in Holland.

((•))

The Dutch train's wheels tapped out a quick tempo. It was a "fast train." A row of tidy townhouses, orderly and well-planned, zipped by. Then they were gone, replaced by lazily turning windmills. Dutch tulips were not yet in bloom, but the windmills alone would have made a tourist's postcard.

A big 747 had delivered me to Holland's busy Schipol International Airport the night before. The KLM flight attendants worked with Dutch efficiency, their blue uniforms offset by blonde hair. They were multilingual, able to converse easily in English, Dutch, French, German, or, in some cases, other languages. The plane's pilot ambled slowly through the aisles, chatting with passengers, resplendent in his tailored, captain's uniform.

After a night at a Schipol hotel, I rose early for my morning meetings. My train to Amsterdam departed from Schipol precisely at the scheduled time, and it arrived at the busy Amsterdam Central Station at the scheduled minute. I boarded another train – this one to Utrecht. Like the first, it was right on time. In Utrecht I switched to a *snel tram* (fast tram) to Nieuwegein, the Utrecht suburb where my colleagues' offices were located.

The WaveLAN group was now part of another company. In September 1996 AT&T had spun off its manufacturing arm into a new enterprise. Unlike the famous 1984 divestiture, in which AT&T was forced to separate from the Baby Bells, this divestiture was the company's idea. It was completely voluntary, intended to help sell equipment to the independent telephone companies that competed with AT&T. All of AT&T's manufacturing businesses, including the Dutch group, became part of the new company. It was called Lucent. The WaveLAN product, the one that had started with NCR and then moved to AT&T, now carried the Lucent name.

I walked toward the Nieuwegein offices, now adorned by Lucent's new circle logo, a red swirl made by a brush dripping with paint. My friend and colleague Bruce Tuch – pronounced "Tuck" – was waiting at the door. We had become friends, often talking by phone. Bruce had visited me at Carnegie Mellon. An American, he was born in New York City – in the Bronx – and grew up in Brooklyn, not far from my New Jersey hometown. Bruce called himself a "real Brooklyn kid."

He had earned an engineering degree at the State University of New York at Stony Brook on Long Island and then moved to the Netherlands, where he worked at Philips Electronics and earned a master's degree at the University of Eindhoven. By the time I met him, Bruce had long been a resident of Holland, living there with his wife and kids. A citizen of both his native and adopted countries, he fluently and effortlessly spoke both their languages.

I didn't speak Dutch, but Bruce and I did have two languages in common, English and radio. We had both started out as radio engineers and understood how radio worked. This distinguished us from engineers and technicians who worked only with computers and data networks. Since Wi-Fi networks depended on radio, and radio was often the weakest link, radio knowledge was a big deal.

The Carnegie Mellon-Lucent collaboration had gone well. At the university we developed wireless design and deployment techniques, and we relayed our experiences and suggestions for product improvement to our friends in Holland. At the same time, Bruce and the other engineers at Lucent worked to develop newer and better wireless LAN products, ones that included new features like the handoff capability that we needed for our network.

Our meetings in Holland covered these technical topics, but I also had something else on my mind. I waited for a one-on-one meeting with Cees Links, the product management director who had, a few years earlier, visited Carnegie Mellon to pitch the WaveLAN product.

Later, in Cees's office I told him that I was ready to expand Wireless Andrew, making it available to everyone at Carnegie Mellon and giving the new wireless service high visibility outside the university. With the new Wi-Fi standard adopted, the time was right for the university to have a truly campus-wide wireless network. We were ready to move from 915 MHz to 2.4 GHz, the radio frequency specified by the new Wi-Fi standard and the one that would be used in the newest WaveLAN products.

Lucent's new 2.4 GHz equipment could help to improve Wireless Andrew. It would allow frequency reuse, and that would improve the network's performance. At the same time, our work could help Lucent guide future customers on how to build networks using the new WaveLAN products. Lucent and Carnegie Mellon could help each other.

But I would need resources – either money or equipment. We had spent our National Science Foundation grant money building the 915 MHz network. I didn't necessarily need money from Lucent. Equipment would be fine. I hoped that Lucent would be able to donate hundreds of their new access points to Carnegie Mellon. If they could do that, I could find money to cover labor and other costs.

Cees gave me an honest answer. He told me that he didn't have authority to grant my request. He suggested that I talk with George Zysman, a Lucent executive in New Jersey. Cees said he would support my request to George, who had the power to authorize Lucent's donation of the equipment we needed. But George would have to be convinced that it was a good thing for Lucent to do.

It wasn't exactly the answer I had hoped for, but there was hope that Lucent might give us the equipment we needed. I would have to go to New Jersey.

(|•|)

A snowstorm raged when my US Air flight from Pittsburgh approached Newark Liberty Airport. In the blowing snow, visibility was bad for the landing. We came down hard.

I rented a car, and, by the time I left the airport parking lot, the snow was tapering off and the sky beginning to clear. I navigated the snow-blanketed New Jersey roads and soon saw the highway exit for Caldwell, my old hometown. The highway passed the tall towers of WVNJ, once the "voice of New Jersey," an AM radio station where I had worked as a college student trying to earn tuition money.

New Jersey had changed. New buildings made old landmarks hard to find. Still, the familiar places triggered memories of teenage ham radio days – afternoons working with Bob, Greg or Ray, building radio equipment, and late nights talking to faraway friends from the third-floor ham shack. Greg and Ray still lived in New Jersey.

My dad was buried in a cemetery not far from Caldwell. I thought about all he had done for me, encouraging my early interest in radio, chipping in to buy ham radio gear, and helping me to pay for engineering school. Without his support, my career – and my life – would have been much different.

But I needed to focus on the slippery roads. I was headed to Whippany, New Jersey to meet George Zysman, Chief Technical Officer of Lucent Technologies Wireless Networks Group. His official biography said that he was responsible for "technology planning, overall architecture and standards, with a focus on ... wireless systems."

A secretary escorted me past a sign marking the office of "Dr. George I. Zysman." George had worked for Bell Laboratories in the days when AT&T ruled the telephone industry, and he made major contributions to the development of microwave and cellular telephone technology. He worked on wireless for most of his career.

Because of his high rank within the company, George's office was spacious and impressive. But the man was friendly and down-to-earth, still an engineer. He wasn't in the chain of command of the Utrecht WaveLAN group, but he had "supervisory oversight" of their research and development work. Translated, this meant that he was a key player in the development of the WaveLAN product line. More important to me, he had the power to authorize the donation of hundreds of Lucent's new 2.4 GHz access points to our project.

I told George about the work we had done in building Wireless Andrew. I told him that we had helped Lucent by advising the Utrecht group on ways to improve the WaveLAN product and on ways to design wireless networks. And I told him that, with the adoption of the new standard, Lucent could be on the verge of a dramatic expansion in the market for the WaveLAN product.

According to my pitch, an equipment donation to the university would also be a major public relations boost for Lucent because I would publicize the expanded Carnegie Mellon-Lucent partnership. And Carnegie Mellon would continue to help improve the Lucent WaveLAN product.

George liked my ideas but wanted to talk to others in the company. He wanted to gather more information, he said. He was friendly and low-key – not evasive. He just wanted some time to think about my proposal.

Later, with snow melting on the roads, I drove to the nearby cemetery and visited Dad's grave. Then I drove back toward the airport. When I reached the highway exit for Caldwell, I turned and detoured through my hometown. I wondered about the house where I had grown up. Was it still painted yellow and green? Were there any remnants of my old ham shack? I drove slowly along the street of my boyhood home, and, as I passed the old house, long ago sold to another family and painted blue-gray, I looked up at the third-floor windows of my teenage ham shack and remembered my experiences with ancient vacuum tube radio equipment. It seemed a distant past.

A few weeks later, George agreed to authorize the donation of 400 new WaveLAN II access points. Under our agreement, Carnegie Mellon would provide the money needed to install the equipment and buy the needed wiring and miscellaneous parts, but Lucent would donate the access points. Both Lucent and Carnegie Mellon would benefit. The team at Carnegie Mellon would continue to develop and improve our Wi-Fi network design methods, and Lucent would continue to improve their product, based, in part, on our advice. George and I soon signed a formal letter of understanding calling for Lucent's donation of the access points and specifying the details of our continuing working relationship.

I announced to the Carnegie Mellon community that we would soon make wireless service available everywhere on campus and to everyone on campus. I didn't announce that we still faced some technical challenges.

CHAPTER 7

PRODUCTS OF OUR IMAGINATION

FEBRUARY 2002

Translucent curtains faded into view. Driven by a warm breeze, they billowed over elegant, upholstered chairs sitting before the open window. As I awakened, I smelled blossoms outside but knew it was February. This couldn't be Pittsburgh.

The breeze's fragrance met the scent of pressed linen sheets. Things were becoming clearer. I remembered arriving the night before in Palo Alto, California. I had checked into the hotel late, so tired I barely noticed my plush surroundings as I carried luggage to my room, unpacked, and fell asleep.

In the morning light, I found an upscale boutique hotel. My room had two meticulously upholstered sitting chairs and a love seat resting on a thick carpet with black squares against a white background. Pink roses filled a vase on the small table. This wasn't the Holiday Inn Express.

There was a knock at the door, and coffee and toast arrived. I recalled placing the order the night before. Rubbing my eyes,

I thought, "I have a meeting this morning." But I had time. It was midmorning in Pittsburgh but still early in California. No need to rush the coffee and toast.

As I sipped the aromatic brew, my mind wandered over the past three years. Wireless Andrew had succeeded. I thought about my last meeting with the design team two and a half years earlier.

((•))

The team's work had continued under the fluorescent lights of the windowless conference room buried deep inside Cyert Hall. The coffee pot at the center of the conference table drew us together. Hands cradled cups all around the table.

Lisa and Mark shuffled big building drawings into the order needed to explain the access point locations they had chosen. The drawings were floor plans for the Carnegie Mellon buildings slated to next receive Wireless Andrew service. Ben fidgeted, eager to get started. Chuck called the meeting to order.

Technicians Lisa and Mark had become skilled Wi-Fi network designers, and the access point locations they had chosen needed little discussion. But there was still plenty to do. Our meetings now focused on a new question – how to make best use of the multiple radio channels made possible by Lucent's latest WaveLAN equipment and the new Wi-Fi standard.

With our move to the higher 2.4 GHz band, we had started using three channels – not just the single channel we had used at 915 MHz. Three channels gave us an opportunity to avoid some problems of the past. In the original Wireless Andrew network, with only one channel available, all stations shared that channel. Only one station – an access point or a mobile device – could successfully transmit at a time. Conflicts occurred. Some courtesy – even among inanimate objects – was needed.

To bring order to the way access points and mobile stations shared a radio channel, the Wi-Fi protocol used the "carrier sense multiple access with collision avoidance" technique. Any station hearing others on the channel was required to *defer* – the station could not send until others were finished. Stations transmitted one at a time to avoid interference and confusion.

But the Wi-Fi protocol, though polite, caused problems when two stations operating on the same channel had overlapping coverage areas. The two stations – access points or mobile devices – although they were not trying to communicate with each other – would, if they could hear each other, defer, each waiting for the other to finish. The result could be slow response time and poor performance for all within the two stations' coverage areas.

Suppose that Lisa and Mark were working in two parts of a building and using mobile computers operating on the same radio channel. Even located in different parts of the building and connected to different access points, their two devices still might be able to hear each other. If Mark's device could hear Lisa's, it would defer each time Lisa's transmitted, delaying messages waiting to be sent by Mark. Similarly, if Lisa's device could hear Mark's, it would be unable to send whenever Mark's was transmitting, degrading her service.

But, with the newly available multi-channel equipment, we could improve things. We could reduce congestion by assigning access points to operate on three different radio channels. There would be no interaction between stations using different channels. But there would be the usual interaction between access points and stations using the same channel.

We added an extra step to our design process. In our meetings, after locations had been chosen for all of the access points in a building, we assigned a radio channel to each access point. The challenge was to avoid co-channel coverage overlap – overlapping coverage areas of access points operating on the same radio channel – as much as possible.

When we chose access point locations, we tried to minimize their coverage overlaps. Then we assigned channels carefully so that, where coverage overlaps did exist, they were between access points operating on different channels.

Choosing the access points' channels was tricky, though. We had only three channels to use. We couldn't achieve perfection. We could minimize but not completely eliminate co-channel coverage overlap.

We stared at the drawings spread on the conference table. Looking at the access points' coverage areas, we all tried to think of the channel assignments that would be best. It was a three-dimensional visualization problem – an exercise in mental gymnastics. We were looking at a set of two dimensional drawings, each showing one floor of a building, and we needed to visualize floor two as being directly above floor one and directly below the floor three. The coverage area of an access point was not limited to the floor where it was located. The access point could also have radio coverage on the floors above and below – and maybe on other floors of the building.

The floor plans showed colored coverage areas, with a color representing the coverage area of each access point. A blue access point on floor two might have blue coverage areas on floors one, two and three. A red access point on floor three could have red coverage areas on floors two, three, four, and maybe others. We struggled to visualize the access points' sometimes overlapping coverage areas.

On this day we faced a construction deadline. Time was short. We needed to complete the design quickly. Our concentration was intense, and the discussion spirited, as each member suggested a set of channel assignments that he thought would minimize co-channel coverage overlap. At last, we reached an agreement that ended the meeting.

((•))

In late 1999 I stepped down as Wireless Andrew's leader and as Carnegie Mellon's chief information officer. I wanted to return to my work as a professor. Other projects were waiting.

But my work with Wireless Andrew wasn't finished. Since 1997, when the IEEE 802.11 standard was adopted, the buzz about Wi-Fi networks and Wireless Andrew had increased. My own writing in journals and magazines was one reason. Countless phone calls, e-mails and letter inquiries showed up at our offices. Other universities wanted to build their own Wi-Fi networks, and they wanted to know how to do it. At first we liked the attention, but soon the flood of inquiries overwhelmed us.

In an October 1999 e-mail message, Chuck complained, "The good news is that Alex's [writing] efforts on Wireless Andrew worked. The bad news is that we are now getting many requests for visits to see it and to discuss it with us." We needed an efficient way to handle all the requests.

We decided to ask our office staff to respond to the phone calls and e-mail messages by inviting inquirers to attend monthly briefings describing the Wireless Andrew network, our design methods, and Carnegie Mellon's wireless research.

In June 2000 I showed up for the first of these meetings. It was in the same conference room that had been used by the Wireless Andrew design team. Waiting there were visitors from three universities – Johns Hopkins University in Baltimore, Cornell University in Ithaca, New York, and St. Francis University in Steubenville, Ohio. They all wanted to learn more about Wireless Andrew. And they wanted to learn how to build their own campus-wide wireless networks.

Chuck opened the meeting, introducing all of the attendees and giving an overview of the Wireless Andrew project. Then Chuck introduced the two Dans. Dan Siewiorek described the series of wearable computers he and his students had built and used with Wireless Andrew since I had first talked to his class in 1994. And another professor, Dan Stancil, described the radio propagation peculiarities of Wi-Fi signals. He had invented a new way to distribute Wi-Fi signals throughout a building by sending them through the building's heating and air-conditioning ducts.

Two other researchers, David and Satya, were absent. David Johnson, with his students, had developed network software that allowed mobile devices to seamlessly switch between wireless networks – between a Wi-Fi network and a cellular network, for example. David was packing. He was preparing to leave Carnegie Mellon for Rice University in Houston, Texas, where he had accepted a new position as a professor. Mahadev Satyanarayanan – Satya – had continued his work with distributed file systems, and he was also busy elsewhere.

But members of the Wireless Andrew design team were on hand to describe the details of our design process. Chuck asked me to speak. I quickly sketched our approach to Wi-Fi network design and mentioned some new work I had been doing to improve the design process even more. Then other members of the design team described their work. The meeting ran all day.

Early the same evening, I walked in Pittsburgh's Shadyside district. Big brick houses stood proudly behind manicured yards. Pittsburgh's business people and professionals lived in this neighborhood just a few blocks from Fifth Avenue, where, a century earlier, the city's banking and steel executives had erected great mansions. Shadyside was more diverse than a hundred years ago, but it was still an upscale neighborhood. I turned onto the fashionable Walnut Street.

With its expensive boutiques and gourmet restaurants, the street was the center of Shadyside. On the warm summer evening, it buzzed with shoppers. Others headed for happy hour or an early dinner. I was going to see my new project team.

The team and I were building the first invention to emerge from the Wireless Andrew project. Although our wireless design procedure had worked well for Carnegie Mellon buildings, it was slow and labor intensive. A way was needed to speed things up, and the new invention offered an answer.

I stepped from the summer evening sunlight into a dark stairwell. After climbing a flight of stairs, I reemerged into bright light streaming through second-story windows. From the clatter of the bar and bistro, I heard my name. "Alex!" someone called. Jon and the guys were sitting at a window table. They were eager to share their pitcher of beer and the latest industry gossip. Talk about the invention would have to wait.

From nearby tables, sports chatter, mostly about the Pirates and Penguins, drifted our way. It was the usual Pittsburgh talk. But our conversation was different. We swapped the latest news of technology company IPOs – initial public offerings – mergers and acquisitions, and talk about new products from the leading network equipment company, Cisco Systems. We barely noticed the bistro's clatter or the hum of traffic on Walnut Street below. And we didn't watch the ESPN screen over the bar, flickering to beckon our attention. We were focused on technology and the business of technology. Ours would have been a typical conversation in a Silicon Valley bar – but not in Pittsburgh.

We turned to the new project. I asked the guys about their work, and each gave a quick report. They were making progress. I pulled out my pen and, using the table's paper place mats and napkins, sketched a drawing of the hardware that was my part of the project.

Jon Schlegel and the other guys were recent graduates of the Information Networking Institute, the same institute that Marvin Sirbu and I had started eleven years earlier. As a student, Jon had been a member of the Wireless Andrew design team, helping with design sessions in the windowless conference room. Sometimes, after the meetings, we talked about how wireless networks might be designed in the future, and I was impressed by Jon's insights. Later we brainstormed about improving the Wireless Andrew design method.

The time-consuming design process wasn't a big problem for a university with lots of students eager to help. But companies building Wi-Fi networks would be different. Most would be unwilling to devote the same time and effort. I thought automation could be an answer, and I began to plan a partially automated Wi-Fi network design tool. Then I recruited a team to write the needed software and asked Jon to lead the group. Hardware design and overall project management would be my job.

Jon and the four other members of the software development team – Mitch, Ben, Kunal and Pratik – shared beer and sandwiches with me. They joked and laughed with obvious camaraderie.

We worked throughout the year 2000 and into 2001. By September 2001 we had a working prototype of the system we called "Rollabout." The hardware was a sight. It was a cart I had built with the help of a friend who owned a bicycle shop. Made of aluminum angle, the primitive machine rolled on four bicycle wheels, a pair of big ones in the back and two smaller ones in the front. A laptop computer rode on the cart's shelf. A Wi-Fi designer could push the cart around a building or other space where one or more access points were installed, and the laptop computer, using colors, created a map showing the coverage areas of the installed access points.

Later versions of the Rollabout hardware looked better. They were more like golf carts. Some could be folded into a small package that was easily transported. But in September 2001 we had our first clunky prototype. Jon and the software team had written 40,000 lines of computer code to give designers the features they needed to quickly and efficiently create a Wi-Fi network. Rollabout took the guesswork out of access point placement. It helped to quickly design Wi-Fi networks with good coverage, good reliability and high performance.

The design tool helped to find good access point locations and choose the best channel assignments, ensuring that a new Wi-Fi network would perform well. It could be used in combination with building drawings, but they were not necessary. Although we had drawings for all of our buildings, we knew that other organizations and companies didn't. They would prefer a tool with no need for drawings.

And Rollabout could estimate how coverage areas would change if access points were moved from their initial locations. If a technician used the on-screen cursor to drag-and-drop an access point to a new location, Rollabout created a new coverage map, based on an estimate of what the new coverage patterns would be. It saved untold hours spent moving access points to new locations and remeasuring signals to find the best access point locations.

The Wireless Andrew design team had worked at a conference table, using a manual, trial and error process to try to find the best set of channel assignments. But Rollabout had a better way. The new software found channel assignments using an automatic process based on coverage areas that had been measured. It was Rollabout's answer to those endless hours spent puzzling over drawings in the windowless conference room.

Wi-Fi channel assignment is related to the famous "four-color map" problem. Mathematicians know that, in general, only four colors are needed to shade the regions of any two-dimensional map and guarantee that no two adjacent regions have the same color. Two regions are considered adjacent only if they share a border – not just a point. For example, at the Four Corners in the southwest United States, Utah and Arizona are adjacent because they share a border, but Utah and New Mexico are not adjacent because they share only one point.

It had long been known that four colors were sufficient to color any two-dimensional map, but mathematicians still wanted a proof. And the "four-color theorem" was finally proven in 1976 by Kenneth Appel and Wolfgang Haken. Other simpler proofs soon followed.

In the Wi-Fi channel assignment problem, access points' coverage areas are "regions" and radio channels are "colors." "Adjacency" is similar to coverage overlap between access points. But coverage overlap, unlike adjacency, is not absolute. Two regions of a map are either adjacent or they're not. With wireless networks, access point coverage areas can overlap a little or a lot – from a few square feet to a nearly complete overlap of many square feet. The Rollabout software found which channel should be assigned to each access point by minimizing the coverage overlap between access points using the same channel.

But Rollabout didn't directly apply the four color map theorem to channel assignment. Most Wi-Fi networks in North America used three channels – not four. And worse, Wi-Fi networks in multi-story buildings were three-dimensional. The four-color theorem's two-dimensional assumption meant that it couldn't be used with a building's three-dimensional wireless network.

So we took a different approach. The Rollabout channel assignment software calculated coverage overlaps from the coverage maps it had created. Then the software assigned channels to access points and calculated the total of the coverage overlaps between access points on the same channel. If two access points used the same channel, their coverage overlap would be added to the co-channel coverage overlap total.

The software tried many different channel assignment combinations, searching until it found the one that minimized the total overlap. The software worked well, and it worked quickly. It was a big improvement over those long conference room sessions.

Rollabout was later patented by Carnegie Mellon and licensed to a Pittsburgh company that sold it as a commercial product. Then it was picked up by a second company and adapted to mapping wireless coverage areas in underground mines. Regulators had decided, after some bad accidents, that Wi-Fi coverage was critical to improving mine safety. To help design these underground networks, beefed-up Rollabout carts moved through the mines, measuring wireless coverage.

((•))

I left the Palo Alto hotel, walking under the big canopy that sheltered the main entry, and crossed the hotel's circular driveway. Uniformed bellmen hustled in the morning sunshine, fetching cars and carrying luggage. Spotting my new colleagues in a car at the edge of the circle, I walked toward them. Bob was at the wheel, and Paul sat beside him. I jumped into the back seat, and we headed to the Cambridge Street offices of Storm Ventures, the venture capital firm that had invited me to Palo Alto. They had decided to invest in a new wireless company.

As we waited in traffic, I talked with Bob and Paul. They were two of the company's first employees, and Bob was one of its founders. They told me about the work they had started and asked me questions about my Wireless Andrew and Rollabout journal papers. There were surprising similarities between my work and theirs.

The new venture capital-funded company would later be called Airespace, and Bob Friday was its chief scientist. He looked the part. With a big beard and abundant hair, Bob might have been a child of the 60s or a very serious and experienced engineer. He was easy-going and unpretentious but full of ideas about the new product. And Bob was a veteran of wireless technology. He had worked at an earlier start-up company called Metricom. That company's "Ricochet" product was used to provide citywide wireless Internet service. The name suggested radio signals bouncing from one base station to another. But the signals were really relayed – received by each station and then retransmitted to the next.

Paul Dietrich was young and slender. His shaved head contrasted with Bob's. Armed with a Ph.D. and experience at three other start-up companies, Paul was ready to continue in the Silicon Valley fray. He was qualified to be a university professor, but, for him, a university job just couldn't compete with the excitement of another new company. And Paul was undaunted by technical challenges.

All three of us had been working to speed up Wi-Fi networks, which were still slower than wired networks. I had been thinking about ways to extend the Wireless Andrew and Rollabout projects to create faster Wi-Fi networks – ones that could adapt themselves to a changing environment, automatically adjusting access points' coverage areas and channel assignments as needed.

The design methods of Wireless Andrew and Rollabout were aimed at the creation of a *static* Wi-Fi network – a network that, once built, uses fixed access point coverage areas and channel assignments, never to be changed until the Wi-Fi network is redesigned.

But Wi-Fi networks operate in a *dynamic* environment. Radio propagation conditions change as objects, large and small – even human beings – move within the network's space. A newly arrived metal object, with its ability to block and reflect Wi-Fi signals, can dramatically change access points' coverage areas. The network's traffic environment changes when people stop and start using their computers or when they switch between applications, reshaping the network's traffic load. I had thought about a new kind of dynamic Wi-Fi network that could adapt to these changes, providing faster service to computer users. Yet I didn't find time to work on the idea. Wireless Andrew and Rollabout kept me too busy.

But the dormant ideas flooded back when I received a phone call from Tim Danford at the venture capital firm. Tim told me that he had been talking with four guys, Bob Friday and three others, about starting the company that would become Airespace. The four wanted to build Wi-Fi networks that could adapt to change. And they wanted to make Wi-Fi networks more secure, too.

About a year earlier, computer scientists at the University of California Berkeley had hacked into a wireless network protected by the Wi-Fi security method of the day. They had shown that Wi-Fi security was weak. Corporate network managers across America didn't like the news, and most refused to build wireless networks that might jeopardize their employers' information. But the four guys with a new company had ideas about how to fix this and other problems.

The partners at Storm Ventures thought the ideas were good ones, and they thought the market was ready for them. When he called me, Tim Danford asked if I'd like to be part of the new venture.

Tim was waiting for us at the door of the Storm Ventures offices. After some small talk about Palo Alto and my trip from Pittsburgh, Tim invited me into his office and closed the door.

It didn't take us long to agree that I would join the new company – but only as a part-time consultant. I also wanted to continue my work at Carnegie Mellon.

Bob, Paul and I were the first members of the company's *radio resource management* team. This meant that, in the coming months, we would work on ways to automatically adjust access points' coverage areas and channel assignments to adapt to a changing radio environment. And we would work on ways to control mobile computers' connections to the access points. The equipment we designed would improve network performance by making the most of available radio resources – the radio spectrum and transmitter power that were available under FCC rules. Bob was our team leader.

In the months that followed, the little company created a new product – one that really revved up Wi-Fi network performance. Users of computers would be able to receive faster responses and send more data through the new wireless systems we developed.

The new Airespace product automatically chose the best channel assignments for access points, using a method similar to Rollabout's. But, in the Airespace product, access points listened to each other, measuring each other's signal strengths and automatically estimating coverage overlaps with neighboring access points. Then, like Rollabout, the new product's software chose channel assignments that minimized co-channel coverage overlap. The Rollabout and Wireless Andrew design methods had chosen the best static channel assignments, but the Airespace channel assignment method was dynamic and automatic.

Channel assignments were updated frequently and with no human intervention. Moving a large metal object like a file cabinet or cubicle partition could make a big change in the radio environment, causing a large coverage overlap between two access points using the same channel. If that happened, the Airespace product reassessed the situation and switched access points to new channels. It was *dynamic channel assignment.*

And the Airespace product controlled the sizes of the coverage areas. By listening to each other, access points detected their coverage overlaps and then adjusted their transmit powers to achieve coverage areas and overlaps that would give the best network performance. The power levels were adjusted automatically without human involvement. It was *automatic cell size control.* Combined with dynamic channel assignment, it simplified the task of designing a Wi-Fi network.

And there was more. In a conventional Wi-Fi network, an access point can become overloaded when it tries to serve too many mobile computers at the same time. But Airespace had a solution – *load-balancing*. Sometimes mobile computers were within range of two or more access points. When this happened, the Airespace equipment automatically guided client computers to lightly used access points so that no single access point would be overloaded. With this scheme, mobiles were more evenly distributed across access points. The load-balancing technique was particularly helpful in heavily loaded networks, and the result was better – faster and more reliable – performance for mobile computers.

And the new product also improved network security, addressing the concerns of those nervous corporate managers. A team developed features to work within a new security structure that had been added to the Wi-Fi standards. The capabilities of the Airespace product, combined with the new standards, convinced some corporate network managers that

Wi-Fi networks could be secure, manageable, and easy to design and build.

Airespace would be sold a few years later to Cisco Systems. When that happened, the Airespace products that Bob, Paul, and others worked on became part of the Cisco line of wireless products.

But, when I first visited the Storm Ventures offices in Palo Alto, the creation of a successful new Wi-Fi product was still in the future. At that moment, Bob, Paul and I could sense that we were on the verge of something big. We thought we could really step up the performance of Wi-Fi networks.

It was time for lunch. We headed to a nearby Mexican restaurant – just the place to start working on some details. Munching on chips and salsa, we pulled out our pens and mechanical pencils and used the restaurant's paper napkins and place mats to sketch the future.

CHAPTER 8

WI-FI FINDS THE FRONTIER

JULY 2010

Back in the small town. It could be anywhere in the American Midwest. There are no traffic lights in the town center – only a flashing red light at the main intersection, where a farm tractor halts and then crawls forward. Railroad tracks divide the hamlet into halves. The original town site sprawls on one side of the tracks, with great green spaces, churches, museums, small houses, and a library. Across the tracks – just past the train depot – shops, offices, more small houses, and the post office cluster together. Invisible radio waves swirl everywhere.

I walk along East Fireweed Avenue in the original town site. Several cars stand in the Catholic and Lutheran church parking lots that face each other across the street. The cars await their owners inside attending committee meetings. Spacious green lawns encircle both buildings.

And the nearby Presbyterian Church's log construction reflects the town's early days. A bottom tier of horizontal logs is topped by others standing on end and embracing the church's windows. More horizontal logs lie atop the windows.

Farther along East Fireweed, I cross South Chugach Street and pass the Pioneer Home. Set behind another big lawn, the home continues its tradition of providing a warm, clean, and comfortable place for our "pioneers," the senior citizens we honor and care for. By a traditional definition, a pioneer has lived in our state for at least 30 years. The Pioneer Home system was created to shelter these old-timers.

A block over, along East Elmwood Avenue, stands the big borough building, the hub of municipal government in our valley. It dominates the surrounding green space. Players on nearby tennis courts focus on their serves. And, on the next door baseball diamond, Little League moms and dads cheer their young players.

The old train depot, perched beside the tracks, is now a community center. The ladies of the quilters' guild are having an all-day meeting there, and the ladies talk as they sew. Their voices can be heard from the depot's parking lot filled with their cars.

On the commercial side of the tracks, just past the train depot, things are a little busier – but not much. Cars move slowly along Main Street, drivers taking special care when children are nearby. People drift in and out of shops, stopping to chat with friends and neighbors. A customer browses the bookshelves at Fireside Books, the little bookstore that beats the odds by daring to compete with Barnes & Noble. The bookstore's owner steers the customer toward new books she might like.

The little town is called Palmer. It sits at the edge of Alaska's wilderness, but without the nearby mountains, it would look like a Midwestern farm town. Meg and I returned to our favorite state – Alaska – when my work at Carnegie Mellon was finished. We settled in Palmer.

We are enclosed by the terrain – snowcapped mountains on three sides and the turbid Cook Inlet, with its prodigious tidal variations, on the fourth. Bald eagles soar overhead. Big moose browse their way through the nearby forest, occasionally feasting on our vegetable garden and ornamental plants. Sometimes a red fox makes an appearance, or a black bear causes a stir. The little burg seems to have been transported thousands of miles – from the prairie of Middle America to the panorama that is Alaska.

Palmer's Midwest flavor is real. It was settled 75 years ago – in 1935 – by 204 families from Minnesota, Wisconsin and Michigan. The families were refugees, driven north by the double whammy of the Depression and the Dust Bowl Era. They settled in a fertile area of south-central Alaska, its promising soils a lucky result of geology and weather. These intrepid souls came north because of Franklin D. Roosevelt's New Deal. Farmer resettlement programs were a rural part of FDR's Depression era plans, but relocating 204 families to Alaska was an extreme case. It moved them to the far end of the continent.

The federal government used two trains to carry the farmers and their families to the West Coast, some to Seattle and some to San Francisco. There they boarded two ships bound for Alaska, eventually arriving in Palmer, then a tent city. The tents were lined up in rows, supported by wooden frames that had been hastily constructed by federal workers just before the families showed up. Living in tents might have been an interesting summer camping experience, but the colonists had expected real houses to be waiting for them.

The government promised to give each farmer a house for his family and 40 acres of land. The farmer was to clear and cultivate the land, raise his crops, and support his family, helping to build a local economy.

But the first winter was rough. The farmers' houses were built slowly by the federal workers, and some of the families spent part of their first Alaska winter still living in tents. The men cleared wooded land to convert it to farmland. There were no chainsaws, and it was backbreaking work. The women struggled to keep house and raise their children in Alaska's harsh conditions. They were determined people, and some stuck it out, but others left, disappointed and discouraged. The Colony Project had a painful start.

Four years later, with only a third of the original families still left in the colony, the government closed down the project. The Alaska market for crops was too small to support all of the families. Ironically, this changed a few years later, when World War II Army troops arrived in Alaska. Then things improved for the remaining colonists, and farming began to thrive in the valley surrounding Palmer. Agriculture prospered for decades, until the late 1960s, when it again tapered off.

But the little town of Palmer has lived on. And many of the colonists' descendents remain. Today Palmer still feels like a farm town, though some of its economic support comes from tourism and from its closeness to the much bigger Anchorage, which Palmer serves as a bedroom community.

Yet Palmer remembers its roots and the hardy families that settled it. Just behind the library rest three monuments listing the names of the original colonist families from Minnesota, Wisconsin and Michigan. The flags of the three states fly beside Alaska's blue and gold banner. And a ship's bell, from one of the federal ships that brought the colonists to Alaska, is on display.

When Meg and I returned to Alaska, we planted ourselves in Palmer. Our Alaskan friends are like no others, and returning to Alaska's tranquility suited both of us after the faster pace of the East.

But wireless has found us again, even in a small town in faraway Alaska. Palmer has 3G cellular service, and Wi-Fi is everywhere. I walk along Main Street – it's really called South Alaska Street – eyeing my smartphone and using it as a radio receiver to look for nearby Wi-Fi access points. It's easy to pick up a Wi-Fi signal. A civic group, whimsically calling itself Radio Free Palmer, has installed five access points to serve the town center, and, remaining true to its name, the group offers the wireless Internet service without charge.

I stop off for coffee at Vagabond Blues, which offers cappuccino, latte, and other espresso drinks to go with pastries and sandwiches. And, not to be outdone by Starbucks, Vagabond Blues has Wi-Fi service. Customers sit at small tables, tapping on laptops, like thousands of others in coffee shops across America and the world.

Back on Alaska Street, I can see several other nearby access points on my smartphone, most of them locked and unavailable to people without the needed password. But one of the access points is in the little office of the Palmer Chamber of Commerce. Because I'm a member, they've given me their password, and I can check my e-mail when I'm near the office but beyond the range of Radio Free Palmer's Wi-Fi service.

I round a building's corner and walk into its parking lot. My e-mail service stops working. The building is blocking Wi-Fi signals coming from the town center. I'm standing in a shadow – not a light shadow but a Wi-Fi shadow cast by the building. The remedy is simple. I step out of the shadow, and my smartphone is again bathed by Wi-Fi waves.

Our little town's Wi-Fi hotspots are beacon lights, pushing their waves in all directions. But, like all Wi-Fi waves, they're weak. They can be blocked by solid objects. They can be reflected, mirror-like, from smooth metallic surfaces. And they can interfere with the waves coming from other Wi-Fi beacons in the cafés, offices and shops along South Alaska Street.

Palmer doesn't have many buildings big enough to have needed the complete Wireless Andrew design procedure. But there were a few that did. The headquarters building of our local telephone cooperative, the Matanuska Telephone Association, has four stories, including the basement, and its Wi-Fi service covers the entire building. The technician who supervised the company's Wi-Fi installation tells me that the access points' locations and their channel assignments were chosen using the design principles that originally came from Wireless Andrew. The same is likely true of the other big Wi-Fi installations in Palmer.

The Wi-Fi equipment installed in Palmer has handoff and other features that Lucent and its competitors developed in the 1990s. But today's Wi-Fi gear has capabilities that go well beyond what we had then. Some of the equipment has enhancements like the dynamic channel assignment, automatic cell size adjustment, and load balancing developed in the 2000s by Airespace and a few other companies.

With electronic assistance, I can sense the radio signals around me. I see them appearing and disappearing on my smartphone. Most people notice that a smartphone is either working or not working. But I know that, when I walk around a building's corner, I will be shadowed from the Wi-Fi signals of Radio Free Palmer. I know that I'll need to find another access point to use – maybe the one in the Chamber of Commerce office. If that fails, I can switch to 3G cellular service.

Usually, my smartphone makes the switch automatically. It has the same ability that was developed by David Johnson in the mid-1990s. I remember what David said in his office the day I asked him to build some software to enable seamless switching between networks. Sitting amid paper stacked high around him, David said, "Oh sure. We can do that." And that's what happened.

With the help of his students, David created the network software needed to work in a multi-network wireless environment, one where wireless communication links can be quickly established and then broken. Today's network software is a distant relative of the software created by David, his students, and their colleagues at other universities. It's used across the United States and much of the world.

As I walk along Alaska Street, the smartphone in my pocket connects to wireless links that start strong but fade when signals weaken. The device scans the radio spectrum for other Wi-Fi and 3G cellular signals, making a new connection if possible. The smartphone prefers a Wi-Fi connection because of its higher speed, but, if no Wi-Fi connections are available, it will settle for a 3G cellular link.

The smartphone itself is a distant relative of the small computers that were developed in the 1990s by Dan Sieworiek and his students. Dan's dream was a wearable computer, one that could be worn on a person's body. Today's smartphones would qualify – including the one riding in my pocket.

Dan and I dreamed about a wireless-equipped wearable computer. But wireless was slow – even slower than the dial-up modems of the day. Wireless local-area networks then were little-known networks using a fledgling technology that might one day offer high speeds, but they were not yet ready to provide the kind of service we needed. We persisted. We thought that the devices would become smaller and that wireless service would become faster and more widely available.

When I touch the e-mail icon on my device, it *syncs up* with a faraway e-mail server, updating the files on both machines. My smartphone's e-mail software can work offline when no connection is available. But, when the device is able to establish a connection to an e-mail server, through either a 3G or Wi-Fi network, it syncs up with the server, making sure that copies of e-mail messages are the same on both the smartphone and server. It transmits messages waiting on the smartphone and receives messages not yet downloaded to the device.

And I can use my e-mail account from different computers at different times – the desktop computer in my office, my laptop computer, computers in hotels and offices when I'm traveling, and, as always, my smartphone. Sometimes I access my e-mail server from two or three computers at the same time. In every case, the local computer syncs up with the e-mail server so that the e-mail messages are the same on the machines at both ends of the link. It's the same capability that Satya and his students created at Carnegie Mellon.

Satya built distributed file systems that worked even when wireless links were unreliable, sometimes connected and sometimes disconnected. It was a challenge to keep consistency among all the files residing on computers in a network, but, in spite of the unreliable communication links, he did it.

We are living in the future that some imagined in the early 1990s. Seldom does the present match so closely a future that was envisioned nearly two decades earlier. Working with those at other universities and companies, my Carnegie Mellon University colleagues helped to invent the building blocks needed to create this future – small computers carried in your pocket and the computer software allowing them to link up and provide seamless service while shielding the devices' users from the complexities of the underlying systems. In spite of the bad boys – the strange ways of radio waves – modern wireless systems provide consistent, reliable service.

At least most of the time.

((•))

On June 7, 2010 Apple's famous CEO, Steve Jobs, was speaking at the Moscone Center in San Francisco. The hall bulged with spectators waiting for Jobs to demonstrate Apple's latest smartphone, the iPhone4. But he was having problems downloading some Web pages from the Internet.

"I'm sorry guys. I don't know what's going on," he confessed to the crowd. "Got any suggestions?" It was an embarrassing moment for Apple and its CEO.

About 20 minutes later, Jobs said he'd figured out there were more than 500 devices in the room using Wi-Fi connections and that the network was completely overloaded. "So you guys have a choice. Either turn off your Wi-Fi or I give up. Would you like to see the demos?" he asked the crowd. "Then all you bloggers need to turn off your notebooks. Go ahead. Just shut the lids. I'll wait," he said.

The Wi-Fi network being used in the Moscone Center wasn't designed to serve the more than 500 laptops and smartphones that were in use that day. And the same thing happens regularly at other high-tech conventions and conferences. Hundreds of people with wireless devices try to use them at the same time, but the convention centers' wireless networks strain under the load.

Just a few weeks after the Moscone network problems, the iPhone4 experienced more trouble. Users of the new smartphone were having problems making phone calls. They were often disconnected. It seemed worse when the devices were held in users' hands in a certain way. Customers were puzzled because the iPhone4 was advertised to provide better – not worse – service.

In a later statement, Apple said the problem was related to the iPhone4's antenna, a metal strip that ringed the device. Suggested fixes emerged, all aimed at insulating the antenna from the user's body. Nail polish could be painted on the antenna. The phone could be enclosed in a case made of an insulating material. People were reminded that smartphones still use wireless – in other words, radio. Even with their advanced apps and high-tech features, they still need a reliable radio link – including a functioning antenna – to work their magic.

And the bad boys are still around.

AUTHOR'S NOTE

Those interested in the details of the technology and events described in *Wi-Fi and the Bad Boys of Radio* may also want to read:

W. Lemstra, V. Hayes and J. Groenewegen, *The Innovation Journey of Wi-Fi: The Road to Global Success*, University of Cambridge Press, New York, 2010.

A. Hills, "Smart Wi-Fi," *Scientific American*, vol. 285, no. 10, pp. 86-94, October 2005.

B. O'Hara and A. Petrick, *IEEE 802.11 Handbook: A Designers Companion*, second edition, IEEE, New York, 2005.

A. Hills and B. Friday, "Radio Resource Management in Wireless LANs," *IEEE Communications*, vol. 42, no. 12, pp. S9-S14, December 2004.

A. Hills and J. Schlegel, "Rollabout: A Wireless Design Tool," *IEEE Communications*, vol. 42, no. 2, pp. 132-138, February 2004.

A. Hills, J. Schlegel, and B. Jenkins, "Estimating Signal Strengths in the Design of an Indoor Wireless Network," *IEEE Transactions on Wireless Communications*, vol. 3, no. 1, pp. 17-19, January 2004.

A. Hills, "Large-Scale Wireless LAN Design," *IEEE Communications*, vol. 39, no. 11, pp. 98-104, November 2001.

A. Hills, "Wireless Andrew," *IEEE Spectrum*, vol. 36, no. 6, pp. 49-53, June 1999

A. Hills, "Terrestrial Wireless Networks," *Scientific American*, vol. 278, no. 4, pp. 86-91, April 1998.

A. Hills and D. B. Johnson, "A Wireless Data Network Infrastructure at Carnegie Mellon University," *IEEE Personal Communications*, vol. 3, no. 1, pp. 56-63, February 1996.

CPSIA information can be obtained at www.ICGtesting.com
Printed in the USA
BVOW080234051011

272860BV00002B/2/P